PowerShell Essentials: Learn to Manage Computers On-Premises and in the Cloud

Table of Contents

Overview

More and more, technology professionals are required to work with information from diverse sources and pull that information together to provide meaning and direction. Sometimes we need to manipulate diverse systems and configure them with similar options. Or you might need a tool for managing cloud resources on a platform like Microsoft Azure. PowerShell can be a single invaluable tool for accomplishing all of these tasks and more. Besides its usefulness when working with Microsoft applications and services such as Microsoft Office, SQL Server, SharePoint or Active Directory, the ability to integrate PowerShell into Linux and other operating systems increases the reach for those who understand how to use it.

The purpose of this tutorial is to cover the basics of using PowerShell to not only access information, but also work with tools from which you create data solutions, such as Azure and SQL Server. You will learn about its capabilities when managing computers or data and get hands on experience through the labs and exercises. This is intended as a reference tool so you can continue to learn after the class is over and the labs are written to facilitate this.

The tutorial has been written with Windows and non-Windows professionals in mind. Those who already have PowerShell tools on their Linux systems can benefit and do some of the labs. If you wish to complete all the lab exercises on a Windows system, you can do so by creating a virtual machine (VM) on a cloud service provider such as Microsoft Azure. Instructions on doing so are included with the material.

The hope is that all who use this tutorial, even beginners, will be able to walk away with enough knowledge and experience to simplify their job tasks and provide a foundation for learning even more. Whether you are a Data Professional, Application Developer, IT Support Technician or Linux Administrator you can benefit from PowerShell and what this tutorial provides.

PowerShell is a command-line shell and scripting language used on Windows, Linux and OS X systems. The tools and options available with this platform allow you to perform configuration and administration tasks using the same methods regardless of the operating system version. The script used to disable the network card on a Windows 10 system can also be used to perform the same function on Windows 8 computers.

Computer professionals familiar with command-line tools in the Windows Command Shell (e.g. dir, copy or cls) or those used on UNIX and Linux command shells (e.g. ls, cp or clear) will be able to carry those skills over into the PowerShell environment. The ability to use existing scripting skills and then build on them makes the transition to PowerShell easier. Sophisticated scripts can be constructed that have access to all the features of the .NET framework.

The uniform way in which different Windows systems can be administered with PowerShell and the tools available that support daily administrative tasks are important reasons to learn this language. PowerShell allows administrators to install applications & features, view & add information to Event Logs, change the domain, local & network properties of a computer, modify security settings and create automated tasks. Besides the operating system, commands can be created to configure applications such as SQL Server (**Invoke-SQLCMD –InputFile "C:\Classfiles\SR1\MSDBQuery.sql"**) or features like DirectAccess (**Set-DAClientExperienceConfiguration –PolicyStore "Contoso\DAPolicy1" -UserInterface $True**).

In this module, students will learn about the capabilities of PowerShell and through exercises, will see practical ways to use this tool.

Lesson 1: Compared to Other Scripting Languages

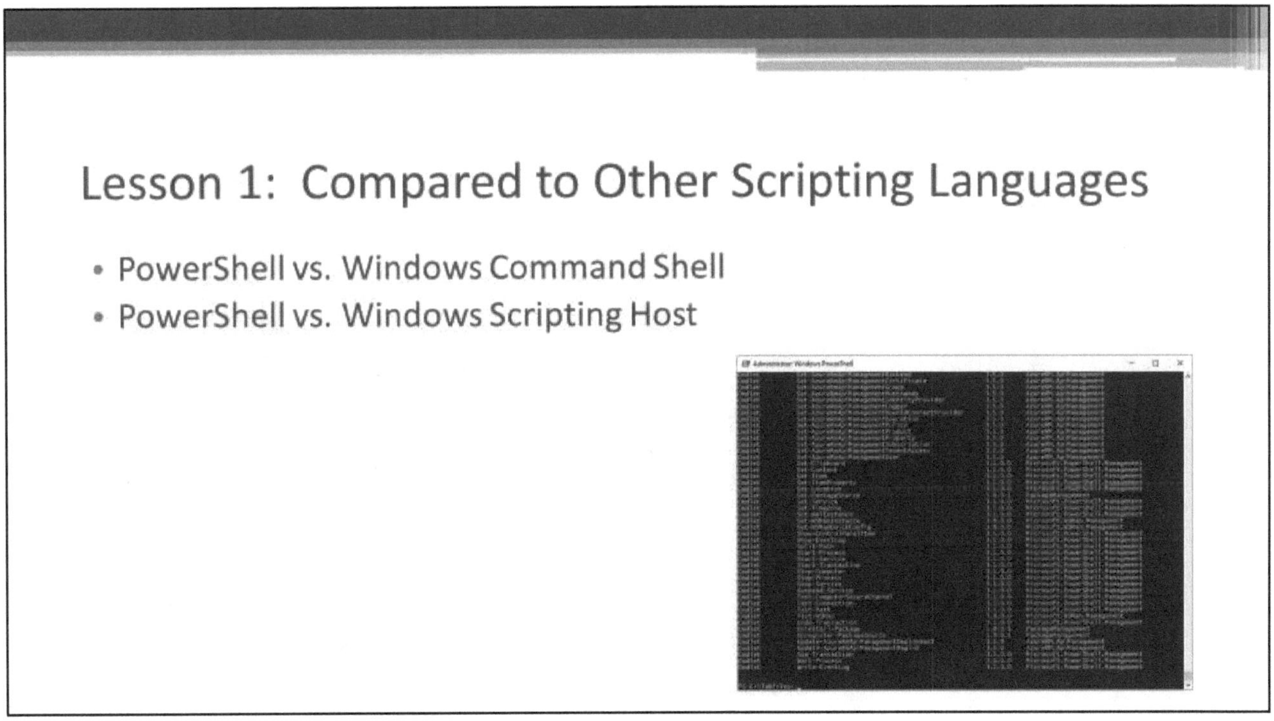

PowerShell features make it the preferred scripting language on Windows devices. Scripts written for popular platforms can be used in it without being modified (CMD and WSH). UNIX and Linux shell commands will also work in a PowerShell environment. It exposes all the capabilities of the .NET framework to the user, so complex tasks can be performed without a lot of programming skills. Administration can be performed remotely over the network and the same commands can be used to administer different Windows versions.

PowerShell will not always be the best tool for a given situation however. Depending on the task being done and the operating system or application that is being used, another scripting language might provide a better solution. Before learning more about the capabilities of PowerShell, we will compare it to other scripting environments such as Windows Command Shell and Windows Scripting Host (WSH).

PowerShell vs. Windows Command Shell

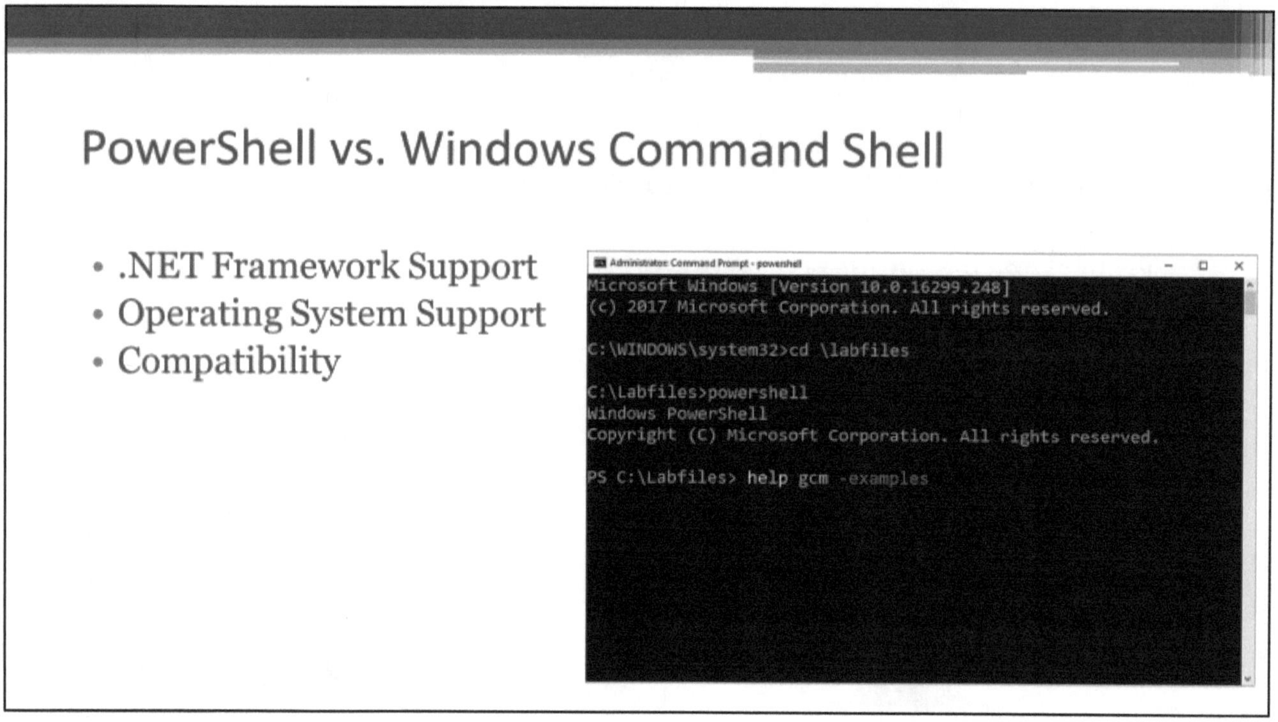

If you are comfortable using the Windows Command Shell (cmd.exe), moving to a PowerShell environment is relatively easy because most of the commands and scripts they used before will still work in this environment. CMD.exe does not have the flexibility of PowerShell and the ability to use the libraries in the .NET framework.

The advantage of using Windows Command Shell is that it will work on all Windows operating systems, not just Windows XP and above. Also, unlike PowerShell, the components are already a part of the operating system and do not need to be installed or enabled.

Now that PowerShell is an open source platform, other operating systems are now introducing it and this trend is expected to grow. PowerShell can now be used on diverse versions of Linux and OS X. While it is not fully functional on these platforms, more features are being added regularly. When specific capabilities are not available for a platform, many take advantage of remoting features which allow you to extend the capabilities of PowerShell beyond what is available on your local computer.

PowerShell vs. Windows Scripting Host

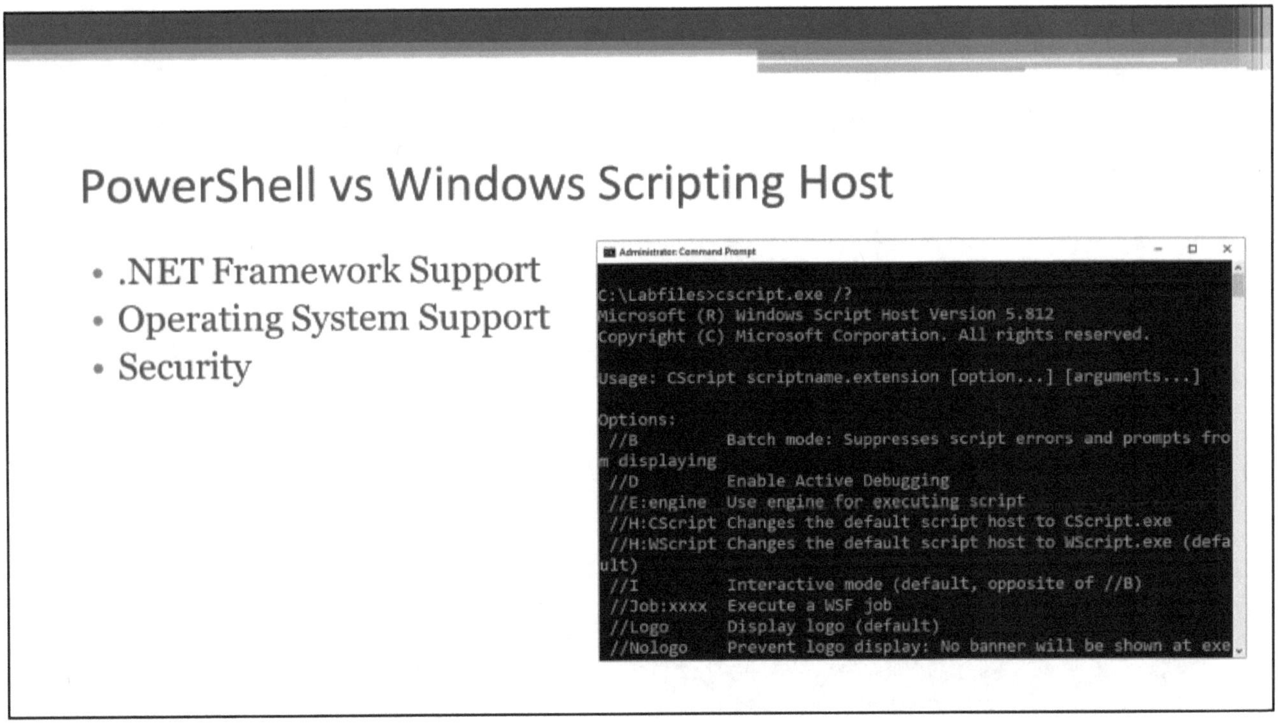

WSH scripts provide the programmer with greater flexibility and access to object configuration than Windows Command Shell. It is available on all Windows operating systems since Windows 98 and can run commands from the command-line or GUI. WSH is a scripting environment and not a scripting language. It requires the use of a scripting language like VBScript or Jscript. It uses the Wscript.exe or Cscript.exe command depending on whether the script is being run from the Windows GUI or Command-line.

Despite its capabilities, many administrators do not use it because the programming environment is not standardized and significant security vulnerabilities are exposed when using it. These issues are overcome with the use of PowerShell.

Demonstration / Exercise 1

Note: Make sure your PowerShell console or ISE has administrator privileges when doing exercises. Any text editor can be used to create or edit the scripts. Students should be encouraged to use a popular PowerShell ISE or another GUI based scripting environment.

Demonstration: Show students how to run PowerShell cmdlets to get user, operating system and computer information. Complete exercise 1 before moving on to lesson 2 or the instructor may decide to complete lessons 1 and 2 together.

Lesson 2: Configuring and Using PowerShell

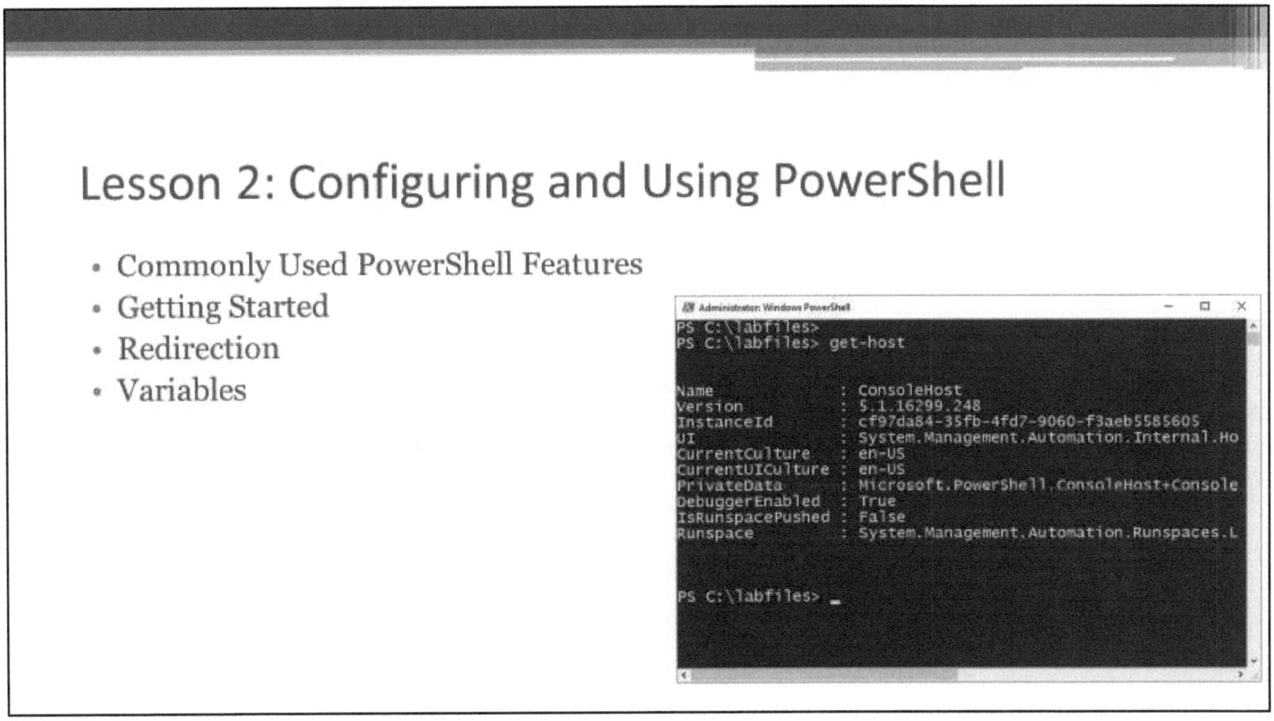

PowerShell is many configurable features and is installed by default on Windows systems. Scripting and other options can be managed per system or configured through group policy. PowerShell is updated regularly so there might be different versions on Windows clients, even if they run the same operating system.

When working on networks with different versions of Windows, it is best to upgrade PowerShell on older platforms so they are all the same. You can verify the version of PowerShell being used on a system by running the command **Get-Host**.

Commonly Used PowerShell Features

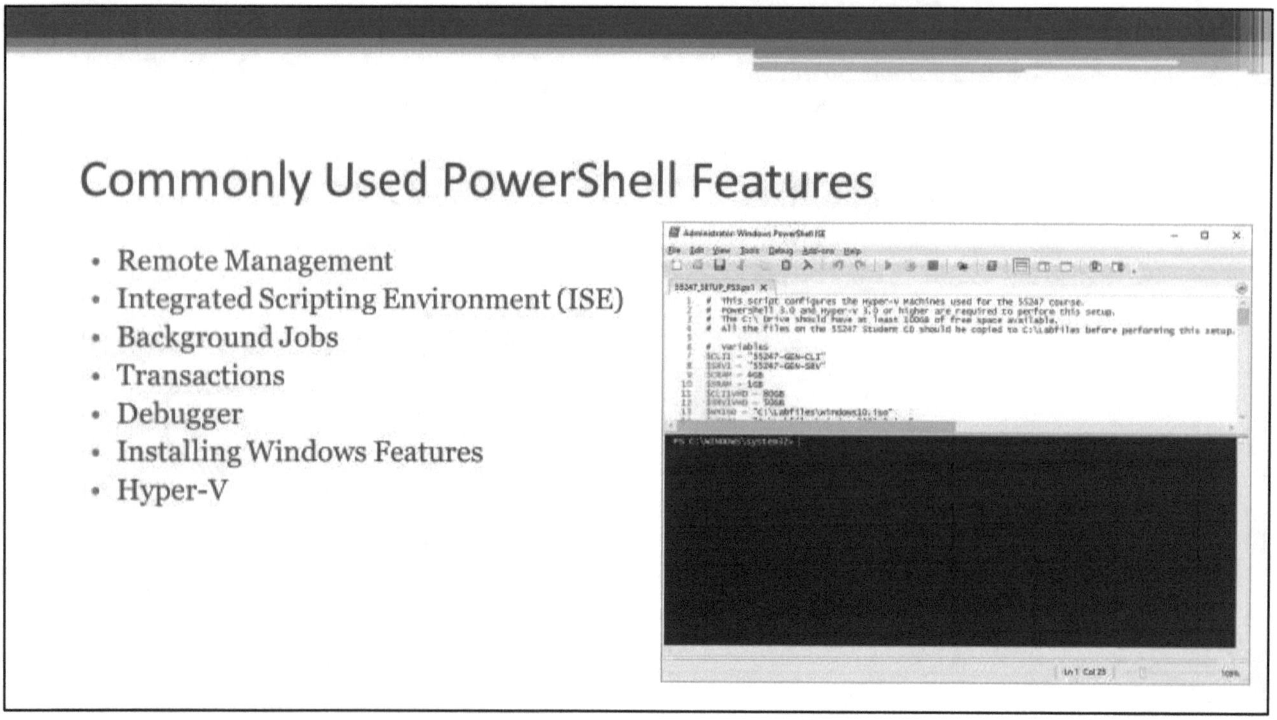

Improvements are constantly being made to the PowerShell environment with the addition of new features and cmdlets. Different applications or resources will sometimes have their own cmdlets, such as SQL Server or Azure. When working with different operating systems, administration will be easier if all of them are configured to use the same version of PowerShell. Some of the more popular features used by administrators are listed and explained below:

- **Remote Management**: One or more computers can be controlled from a single remote system (e.g. **Enter-PSSession -ComputerName Server1**). It is also possible to control virtual machines run in a Hyper-V environment using PowerShell Direct (e.g. **Enter-PSSession -VMName NanoServer01 –Credential Contoso\Administrator**). PowerShell Direct is currently available on Windows 10 and later operating systems.

- **Integrated Scripting Environment (ISE)**: ISE allows technicians to write and debug PowerShell scripts in a GUI environment. Keyboard shortcuts and menu options can now be used to perform options that were once available only from the command-line.

- **Background Jobs**: Jobs, in the form of PowerShell scripts can now be run in the background which allows you to continue working in the shell. (e.g. **Invoke-Command -ComputerName Student10 –File configureAzureVM.ps1 -AsJob -JobName "AzureJob10"**). Scheduled jobs can be created with **Register-ScheduledJob**.

- **Transactions**: Operations can be grouped together to have them complete or fail as a single unit. If a command fails, other commands that already succeeded but belong to the same group can be reversed. (e.g. **Start-Transaction ; .... ; Complete-Transaction**)

- **Debugger**: Scripts and code can be checked for errors. You can step through your code and create breakpoints to find problems more quickly.

- **Installing Windows Features**: Windows programs and features can be installed and uninstalled with PowerShell. Updating existing feature configurations is supported. (e.g. **Get-WindowsOptionalFeature -Online | Select-Object FeatureName, State | Sort-Object FeatureName**). Get-Module can be used to list modules that can add functionality to the PowerShell environment (e.g. **Get-Module –List**)

- **Hyper-V**: If the hardware supports it, Hyper-V can be installed and configured on Windows devices to create virtual networks and machines. Most of the options available on Windows Server are also available on Windows 10. (e.g. **Import-Module Hyper-V ; Get-VM –ComputerName NYC-DC1**). Windows 10 and Windows Server 2016 allow Hyper-V to be installed on virtual machines.

Getting Started

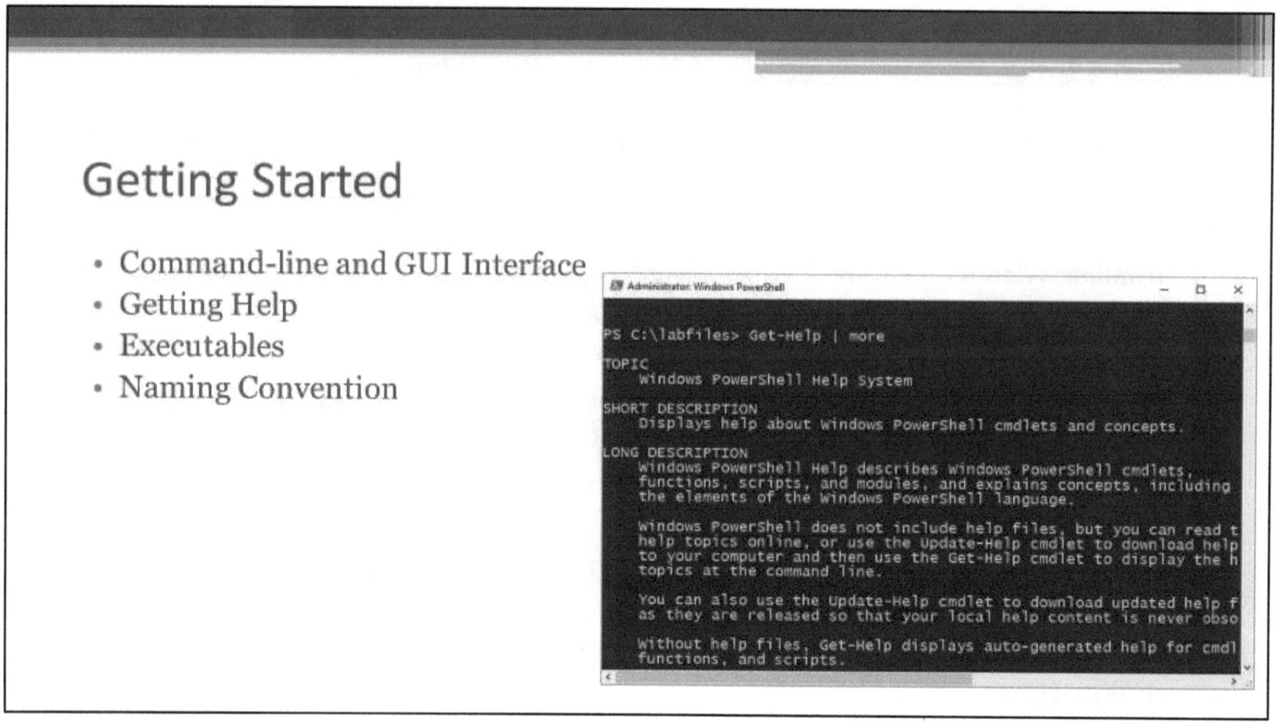

The PowerShell shell can be accessed from the command-line or GUI. Running the command **powershell.exe** will create a command-line shell and **powershell_ise.exe** will open the Graphical User Interface. For those who already know how to run commands from the Command Prompt or WSH, they will be able to use their commands and scripts for the most part.

To start using PowerShell commands, simply type the name of the command and press enter. The cmdlets are not case-sensitive. Cmdlets such as Get-Service are executables that provide preconfigured functionality without having to create the code yourself.

Some cmdlets can be used to help you get started in PowerShell by providing helpful information about existing commands and their syntax. Running the command **Get-Help Get-Service** provides information about how the get-service command can be used. Using the **–Detailed** parameter can provide much more information if needed. Running **get-command** will list all of the preconfigured commands available to you and **get-command set-*** will only list the commands that start with set-. The wildcard characters (* or ?) can be used to add flexibility to the commands that are executed. To get information about all commands with the word computer, you could run the command **get-command *computer***.

Detailed information about a command and how to perform operations in PowerShell can be accessed by using built-in documentation. The command **get-help about** will show a list of the available libraries and individual libraries can be accessed in the same way (e.g. **get-help about_scripts**). Help information for cmdlets can be updated by running **Update-Help**.

Many of the commands used are not cmdlets, but aliases. Commands familiar to those who work with UNIX or CMD.exe will work because they have been mapped to existing cmdlets that perform the same function. The ls, dir, clear and cls aliases are examples of these. To see the cmdlet being used to support an alias, use the command get-command (e.g. **get-command cls**).

Because of the straight forward naming structure of the cmdlets, it is relatively easy to understand what a command will do and to get at the names of related commands. The two-part naming structure starts with a verb and then a noun. Cmdlets that retrieve information will always begin with **get**, those that add information begin with **add** and if existing information is being changed, they will begin with **set**. The second part of the name, the noun, represents the object that is being worked with. So **Get-Eventlog** will refer to the event logs on a computer.

Redirection

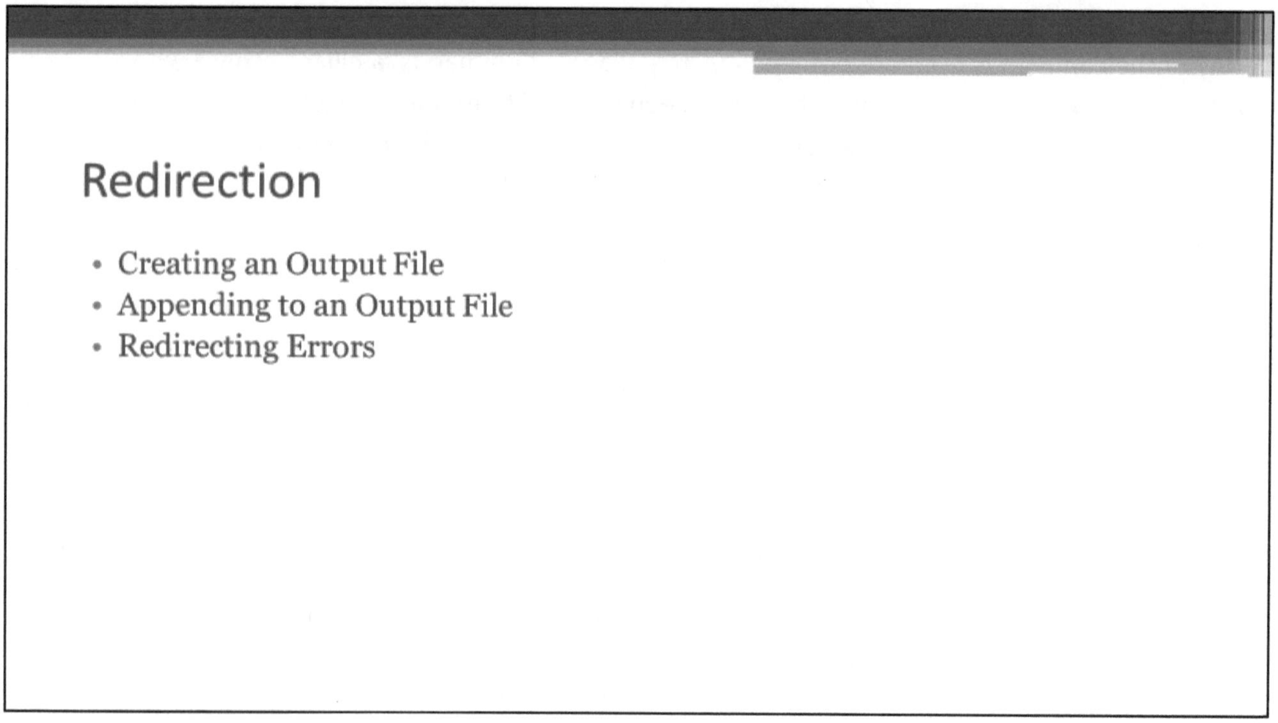

A very useful feature in most scripting languages is the ability to redirect output intended for the computer screen to a different location. PowerShell is also capable of this and this feature is often used to capture information from errors or for logging purposes.

A number of redirection operators are available to meet different needs. The redirector (>) is used to forward information that is normally sent to the screen to a different location, such as a file (e.g. **Get-ChildItem > DirList.txt**). If the file being written to already exists, it will be overwritten with the new information. Two redirectors (>>) are used to append to the file when it already exists. If it does not exist it will be created. If the information being captured was generated from an error, use the number 2 in front of the redirector (2>). To append the information to an existing file instead of creating a new one, use the number 2 and two redirectors (2>>).

Variables

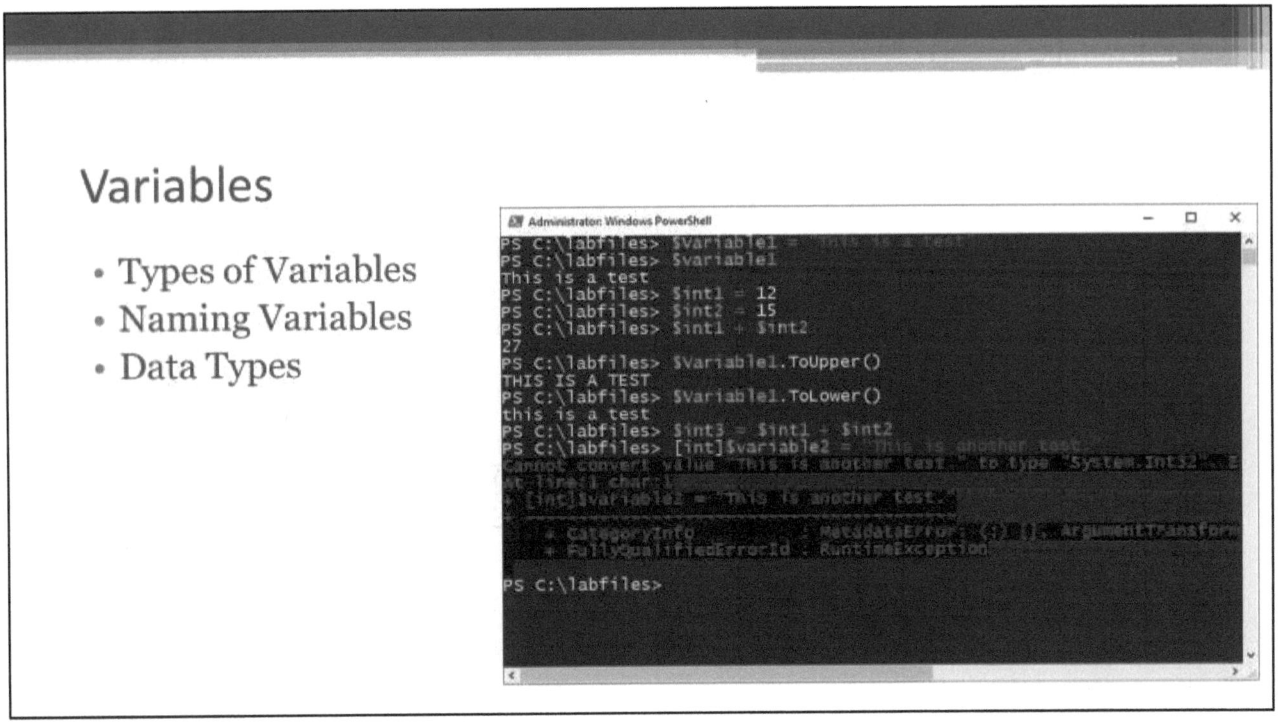

PowerShell uses variables as a holding area for data that is used in different parts of your code. The variable name is defined once, but the information in it can be used as many times as needed. In addition to the built-in variables you can define your own. The variable name always has a prefix of **$** and values can be assigned to them at the time of creation (e.g. $Variable1="This is a test").

The naming conventions and characters allowed are very flexible. Although the use of special characters and spaces is not recommended they can be used if the variable is enclosed in braces (e.g. **${Variable.1}="This is a test"**). To avoid issues with variable names, only use letters of the alphabet, numbers and the underscore character. The name of the variable should also be descriptive of the kind of information it will contain. Community best practices also recommend using camel casing for the names. The first word is in all lower-case and subsequent words have the first letter capitalized (e.g. firstPrintServer).

Variables are most often used in scripts to simply the code and make it more readable. The information they contain can also be manipulated when necessary and the new variable values made available in the script. You can also manage the data type of the variable which allows you to perform specific tasks on them related to the kind of information they store.

For example the use of integer values (**$int1 = 4 ; $int2 = 5 ; $int3 = $int1 + $int2**) allows you to perform mathematical calculations on the variables. Working with text information however (**$string1="This is a**

test "; $string2="of concatenation."; $string3=$string1 + $string2; $string3**) allows the information to be concatenated or manipulated in other ways (e.g. **$string3.ToUpper()).

PowerShell can automatically change the data type of a variable based on the information that is assigned to it. In some cases, you might prefer to hard-code the data type to control the kind of data that can be assigned to a variable. Using the expression **[int]$value1 = 5** for example, would not only assign the number 5 to the $value1 variable, but also generate an error if you tried to change it to a non-numeric value.

Demonstration / Exercise 2

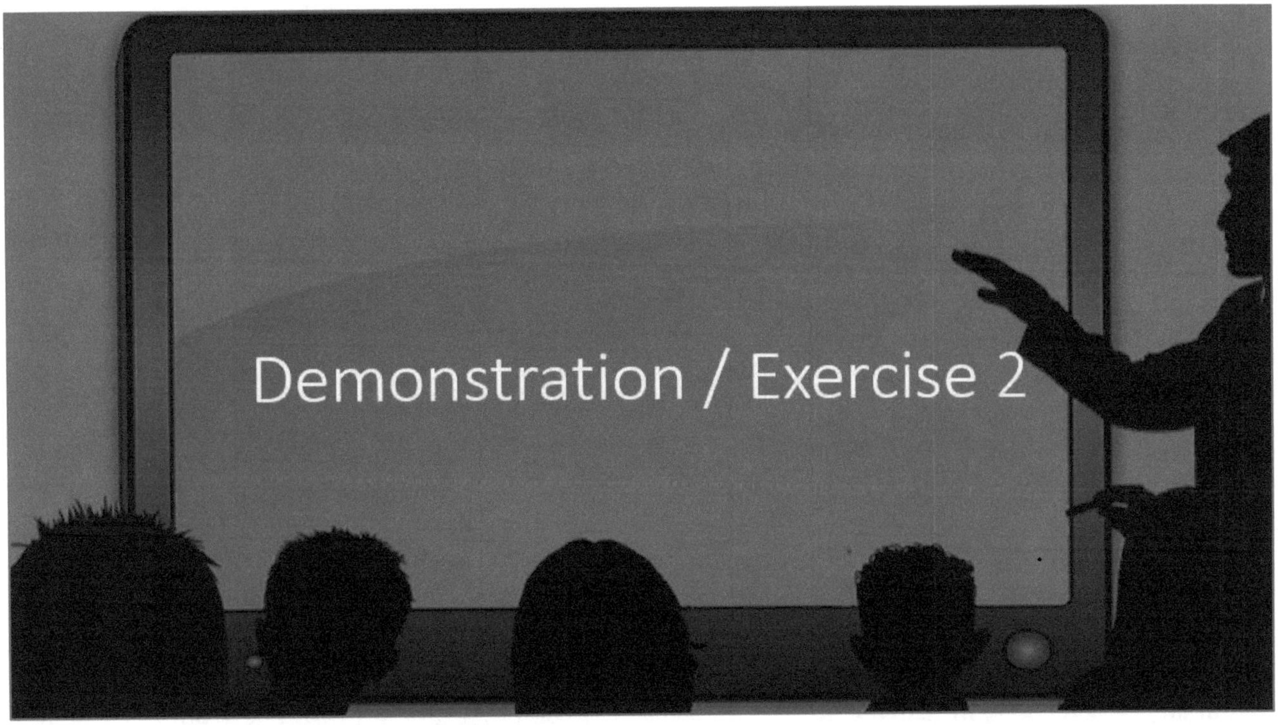

Note: Make sure your PowerShell console or ISE has administrator privileges when doing exercises. Any text editor can be used to create or edit the scripts. Students should be encouraged to use a popular PowerShell ISE or another GUI based scripting environment.

Demonstration: Show students how to get help for PowerShell cmdlets and use built-in documentation to find the options available for a specific cmdlet. Complete exercise 2 before moving on to lesson 3 or the instructor may decide to complete lessons 1 and 2 together before moving onto lesson 3.

Lesson 3: Creating and Running Scripts

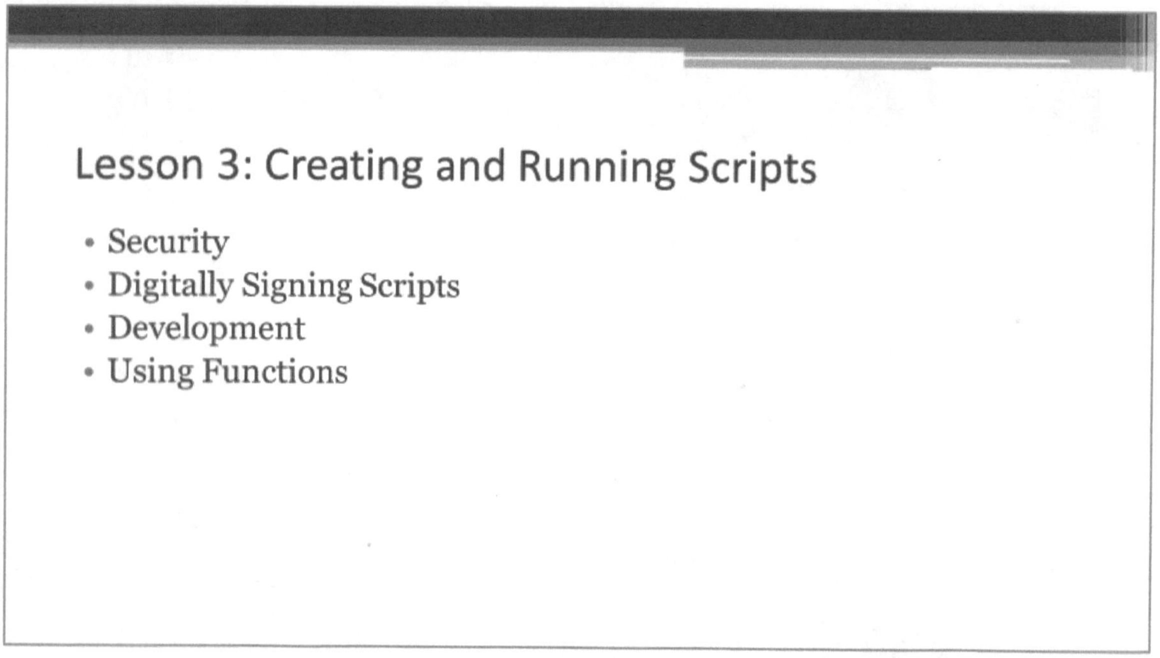

The ability to run code interactively from a shell is a very useful feature of PowerShell, but the scripting capabilities multiply its effectiveness by allowing you to do multiple operations at the same time. This is done by saving your code to a text file with a PS1 extension. This file can then be executed which carries out all the actions the code would have performed interactively. These files can be used for day-to-day tasks, scheduled operations and Windows Troubleshooters.

The only resource needed to create a PowerShell script is Notepad or another text editor. The PowerShell Integrated Scripting Environment (ISE) is very useful because of its GUI interface and debugging features. It is the preferred tool for creating scripts on Windows platforms. For non-Windows systems, there are free tools available with similar functionality.

As with variables, script names should be descriptive and avoid the use of special characters. The cmdlet naming convention which uses a combination Verb-Noun syntax is also recommended (e.g. **Get-Service**). The pascal casing convention which capitalizes the first character in each word used in the name is also recommended. Useful coding examples and tutorials can be found at www.microsoft.com/powershell. Scripts that are used regularly may be broken into separate tasks and defined as functions to make them easier to use in some situations.

A major concern with any scripting language is security. Mechanisms are needed to prevent accidental execution of scripts and measures must be taken to prevent unauthorized code from running. We will see how PowerShell is configured to address these concerns. Writing a script is relatively simple if you already know how to perform an operation interactively with the correct code, but we will also look at recommendations for configuring and securing the scripting environment.

Security

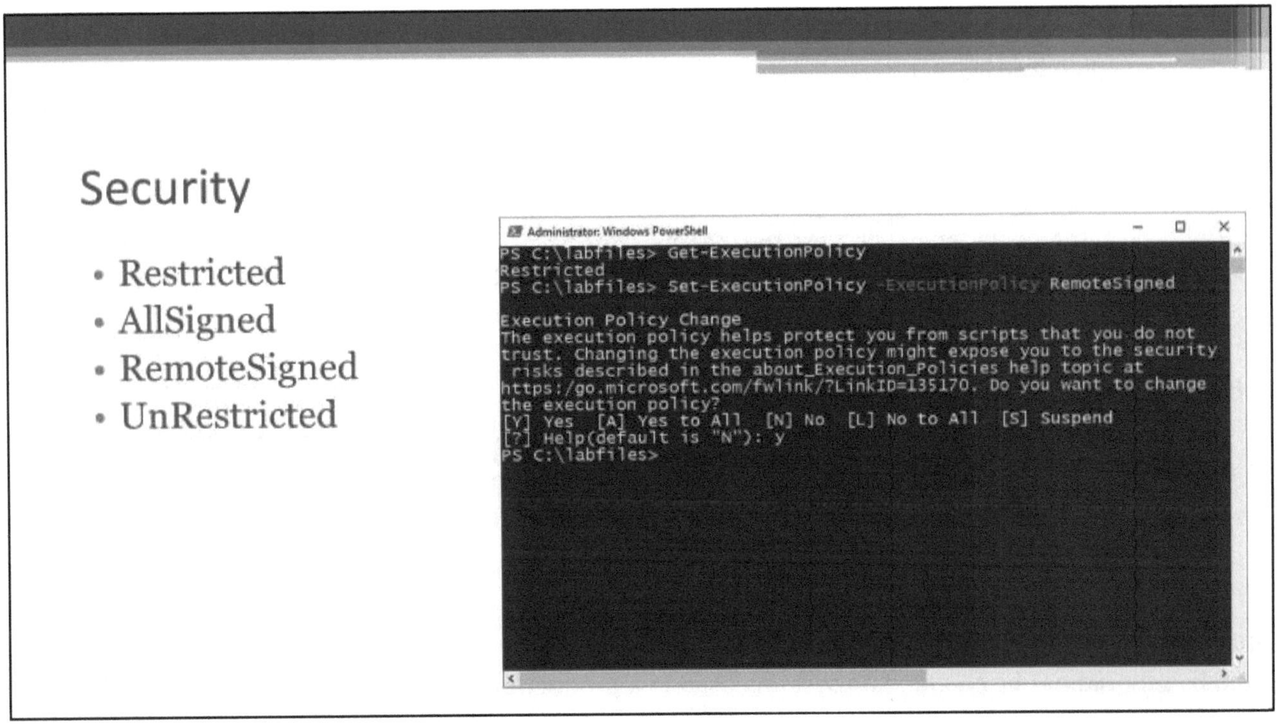

The default configuration of PowerShell prevents the execution of scripts by default. Interactive execution of code is always possible, but the ability to run PS1 files must be enabled. The setting is controlled by configuring the execution policy. It can be set to Restricted, AllSigned, RemoteSigned or UnRestricted (e.g. **Set-ExecutionPolicy AllSigned**). As with other options that affect the security of the system, administrative credentials are necessary to run this command.

- **Restricted**: This is the default and most secure setting. No PowerShell scripts can be executed on the system if this setting is left in place. Everything must be done by running the code directly from the shell. It does allow the execution of configuration files if they are signed and the publisher is trusted.

- **AllSigned**: This option allows scripts to be executed but only if they are digitally signed by a trusted publisher. This is the most secure option that allows the execution of scripts. If the computers are configured for remote management using PowerShell, this requirement will apply to remote scripts as well as local ones.

- **RemoteSigned**: This option allows scripts to be run locally or remotely, but requires that scripts downloaded from the Internet zone be signed by a trusted authority. Because the metadata identifying Internet downloaded files can be easily changed, this should not be considered a secure method to prevent the use of Internet scripts on the local network.

- **UnRestricted**: This option allows all scripts, local and remote, to be executed without being signed. Scripts executed remotely will present a warning message that must be acknowledged before being run.

While the UnRestricted setting might make it easier to execute scripts, it is not recommended for security reasons. The most secure scripting option, AllSigned, can be implemented in a domain environment without extensive changes. PowerShell options, Remote Management settings and Certification deployment can all be configured and maintained through Group Policy. If you still need the ability to run scripts locally without signing them, the RemoteSigned option can be used.

After enabling an appropriate execution policy, scripts still must be executed by specifying the full path. If the script is in the local folder, it can be run by typing ".\" in front of the script name (e.g. **.\TestScript.ps1**). This helps to prevent a common problem in other scripting environments of scripts being executed accidentally or opening the door to the possibility of running malicious code.

A further security measure is that PS1 files are opened by default with notepad.exe and not powershell.exe. Double-clicking a script from File Explorer will simply open it Notepad. This association can, but should not be changed on systems used in a production environment. It's a further mechanism to prevent accidental execution of scripts. Whenever possible, all scripts should be tested and executed without Administrative privileges before deployment to a production environment.

Digitally Signing Scripts

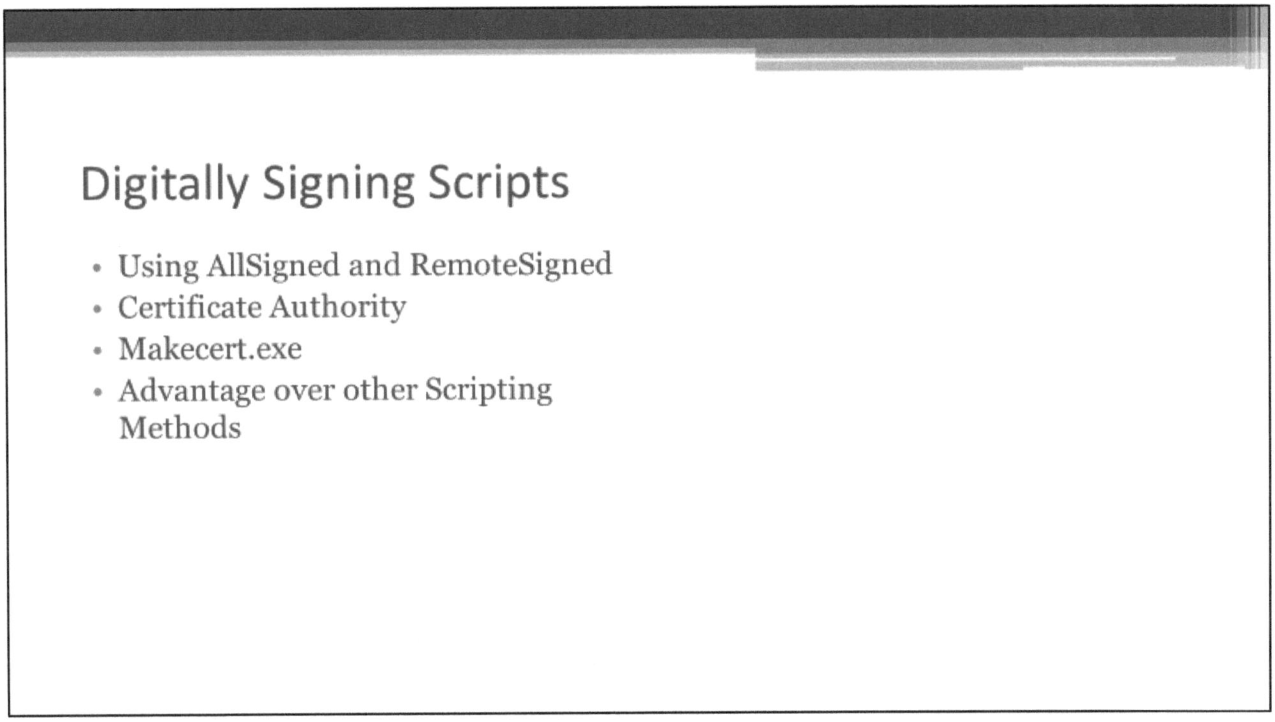

If the recommended execution policies of AllSigned or RemoteSigned are used, there will be a need to use a certificate management system. These certificates are used to "sign" a script and therefore identify its source. The computer system can identify this source from the information in the certificate and compare it with a list of trusted certificate sources that the network administrators can manage through Group Policy. Scripts signed by sources that are not on the trusted list can be prevented from running. This helps to protect the network from malicious code or unapproved scripts.

The server that issues the certificates (Certificate Authority or CA) can be from a trusted Internet organization (e.g. VeriSign). If the additional cost of doing this is not permissible or if the certificates will be used purely for Intranet purposes, an intranet server can be configured to generate and store the certificates.

An easy way to sign scripts is to use the **makecert.exe** tool to create them. It is one of the tools installed with the Windows Software Development Kit. Makecert.exe can generate a new certificate for signing PowerShell scripts without the need of a Certificate Authority. Because this certificate would be generated locally and self-signed, it would need to be added to the domain list of trusted publishers. This can also be done through Group Policy.

While the steps to get the certificate infrastructure working might take some time, it would only have to be done once and would help to secure your scripting environment. This is one of the key advantages of PowerShell vs. other scripting methods, so it should be fully utilized.

Note: The **New-SelfSignedCertificate** cmdlet can be used to create new certificates, but Microsoft documentation encourages its use only for testing purposes.

Development

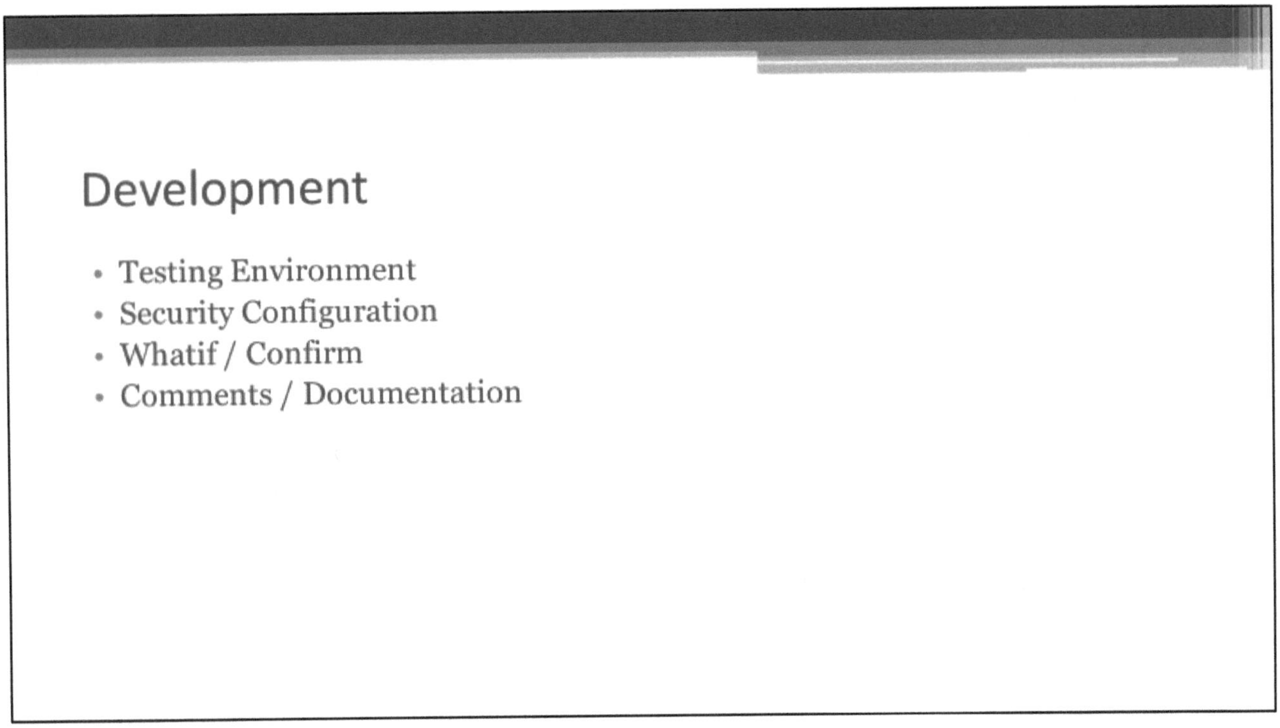

Scripts should always be tested thoroughly before deployment to a production environment. Depending on the amount and level of scripting done, separate development computers or virtual machines might be set aside for these tasks.

Some of the normal security measures might be disabled in such an environment since production systems will not be at risk. Disabling the requirement to sign scripts before execution can speed up development if the right precautions are taken.

If the testing is being done on production systems, all security precautions previously mentioned should be in force. In particular, make sure that digital signatures are required for remote execution and that powershell.exe is not associated with the PS1 files.

When testing the code to be used in a script, use the **WhatIf** and **Confirm** parameters with the cmdlets to make sure that no unintended changes are made. The WhatIf parameter will display a message showing what would happen if the command were executed without actually running it (e.g. **Set-Alias Srv Get-Service –WhatIf**). The Confirm parameter will actually execute the statement but only after the user is given a prompt which allows them to continue or stop the operation (e.g. **Set-Alias Srv Get-Service –Confirm**).

It is also good practice to include comments in a script to make them easier to understand. Using the pound (#) symbol as the first character in a line will let the system know that all the information on that line is to be ignored when interpreting your code. If the script is used for an important operation and extensive comments are needed, an additional text file with the necessary information might be included as a part of the deployment process.

Using Functions

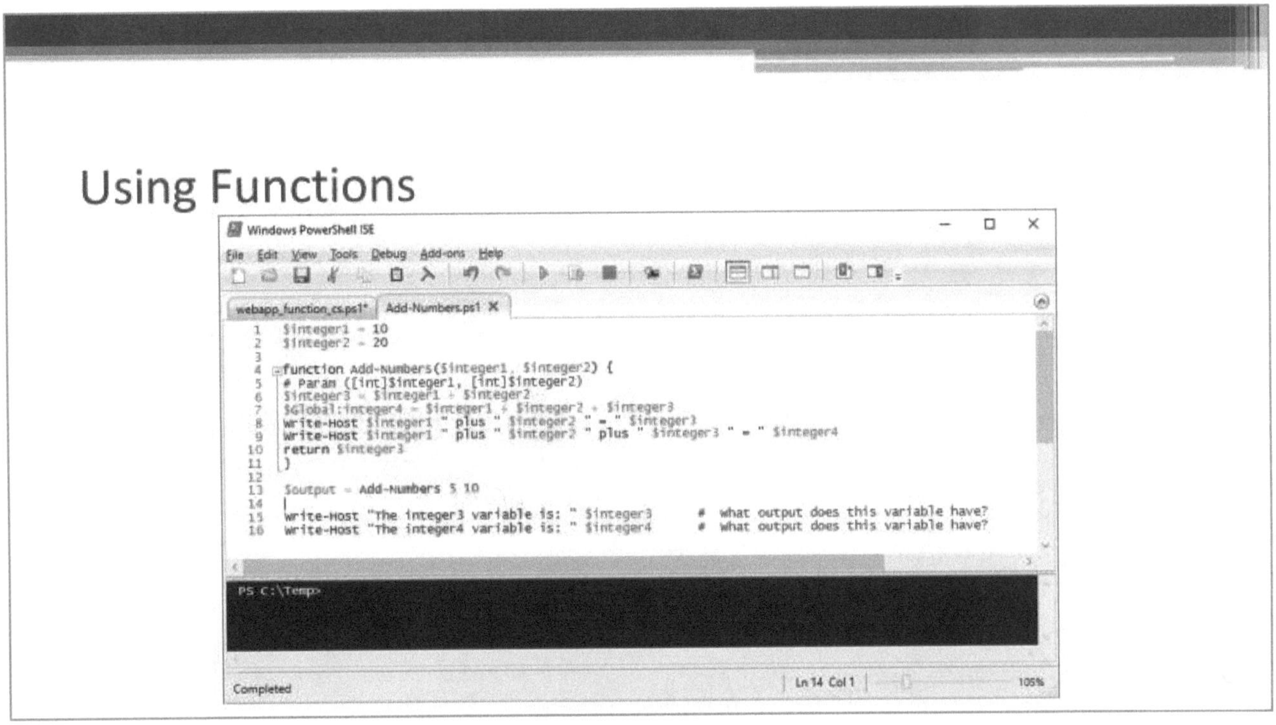

Creating your own functions in PowerShell often makes sense when the same code is being used repeatedly. Defining the task once as a function allows it to be used as many times as needed.
The keyword used to define a function is "function" followed by the name used to call it. The name is followed by parentheses within which you can define input parameters if necessary. Input parameters may be defined before the parentheses using brackets or the keyword "Param" can be used to define them at the beginning of the function definition within the parentheses.

The return statement is used to pass values to the shell after the function is finished running. Another way to pass variable information to the calling shell is to use global variables.

Demonstration / Exercise 3

Note: Make sure your PowerShell console or ISE has administrator privileges when doing exercises. Any text editor can be used to create or edit the scripts. Students should be encouraged to use a popular PowerShell ISE or another GUI based scripting environment.

Demonstration: Show students how to create and execute PowerShell scripts. Use **Get-ExecutionPolicy** and **Set-ExecutionPolicy** to manage the execution settings for scripts. Complete exercise 3 before moving on to lesson 4.

Lesson 4: Administering Local Resources

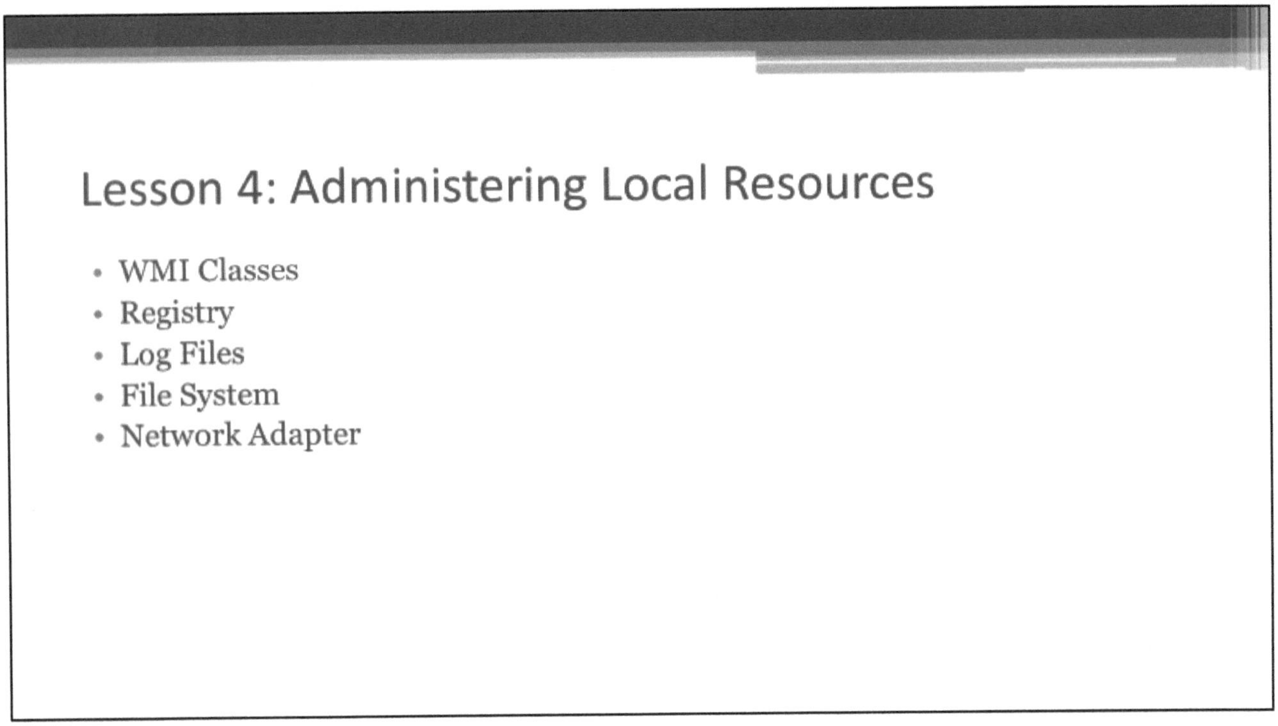

Because PowerShell uses the .NET Framework libraries, it can be used to access and change almost any component on a system. Access to WMI classes makes it relatively easy to administer operating system services and devices. As a technology professional, you can take advantage of this to manage the computer registry, log files, file system and different devices such as the network adapter.

WMI Classes

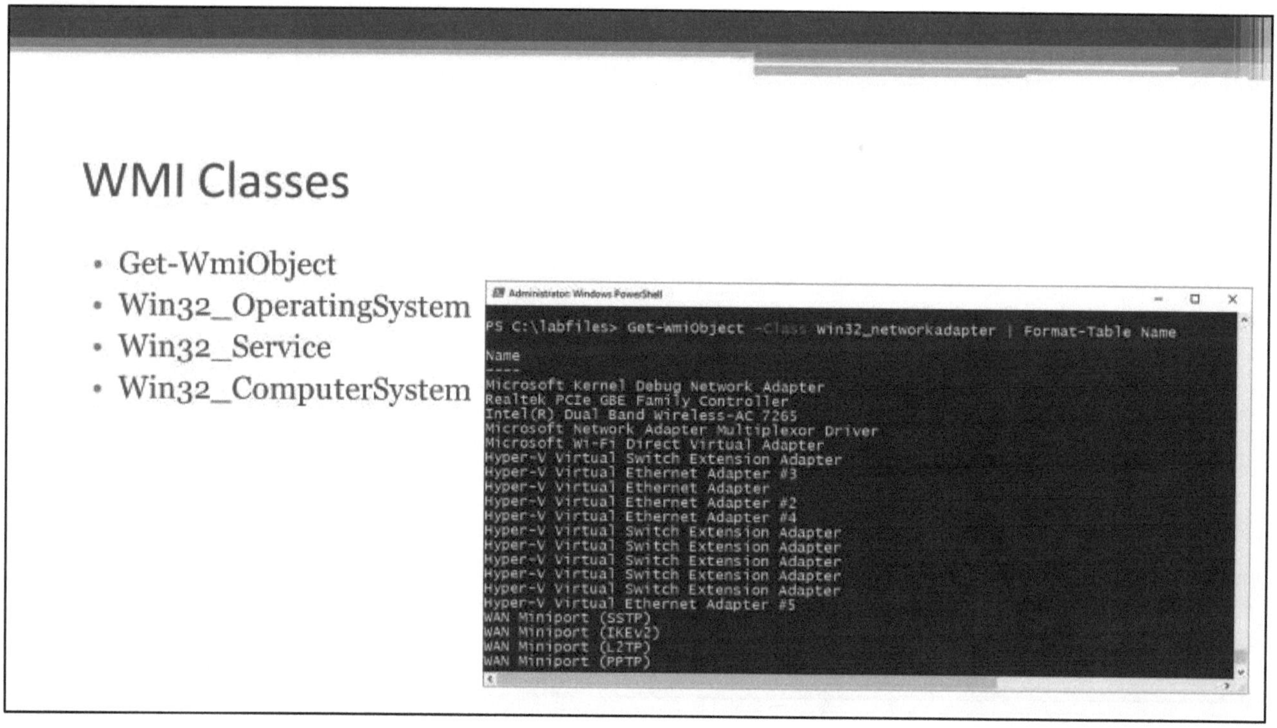

An important path to accessing information about components on a computer is the WMI classes. Using the Get-WmiObject cmdlet as a gateway allows you to view details about important computer components. Information about the computer can be accessed from Win32_ComputerSystem, operating system data can be retrieved with Win32_OperatingSystem. A list of services running on a system can be retrieved from Win32_Service and disk information is available in Win32_LogicalDisk. Network adapter and configuration information can be accessed with Win32_NetworkAdapter or Win32_NetworkAdapterConfiguration (e.g. **Get-WmiObject Win32_ComputerSystem**).

A list of available classes can be viewed by running **Get-WmiObject –list**. The classes available and the information they provide are extensive. Tools such as **PowerShell Scriptomatic** can be used to browse this information graphically.

Registry

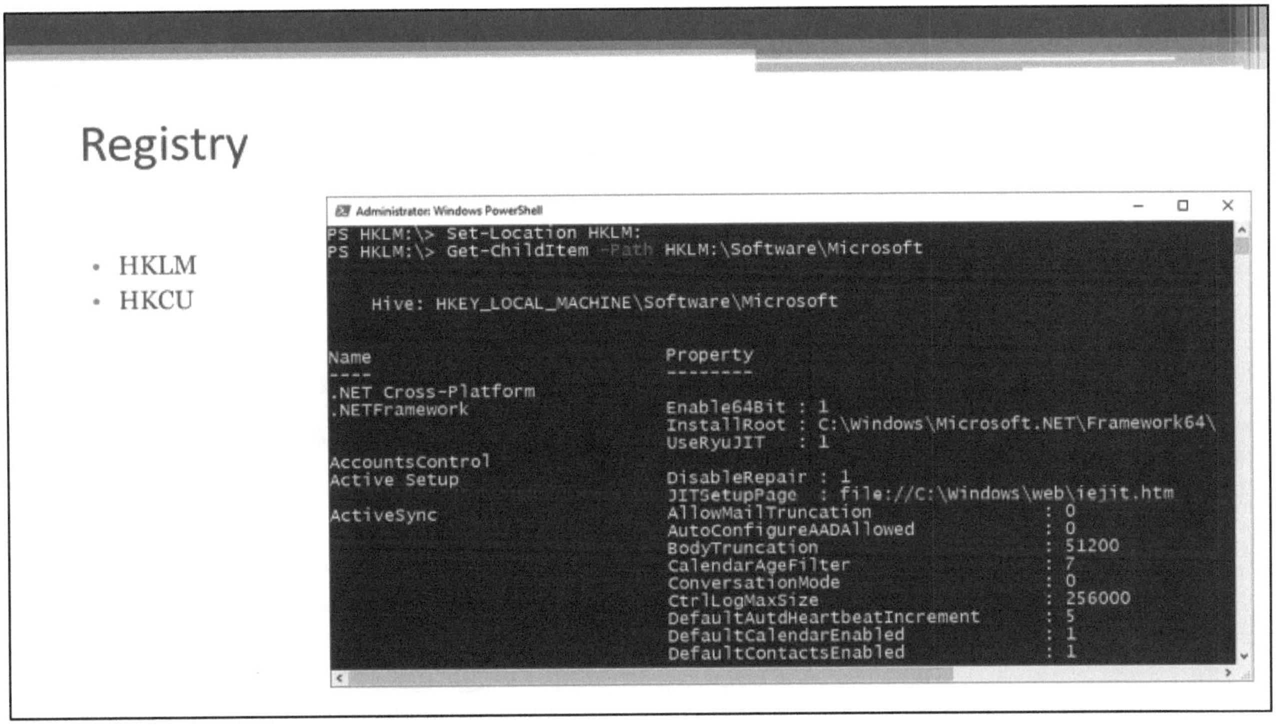

Information in the registry can be accessed and manipulated like data on a file system. Querying information can be done by accessing the right .NET class such as Microsoft.Win32.Registry or Microsoft.Win32.RegistryKey. The Get-ChildItem, New-Item and Remove-Item cmdlets can also be used to browse and change registry information (e.g. **Get-Childitem –Path HKLM:\Software\Microsoft** or **New-Item –Path HKCU:\TestKey**). Set-Location can be used for connecting to and navigate the registry as you would the file system (e.g. **Set-Location HKLM:**).

Log Files

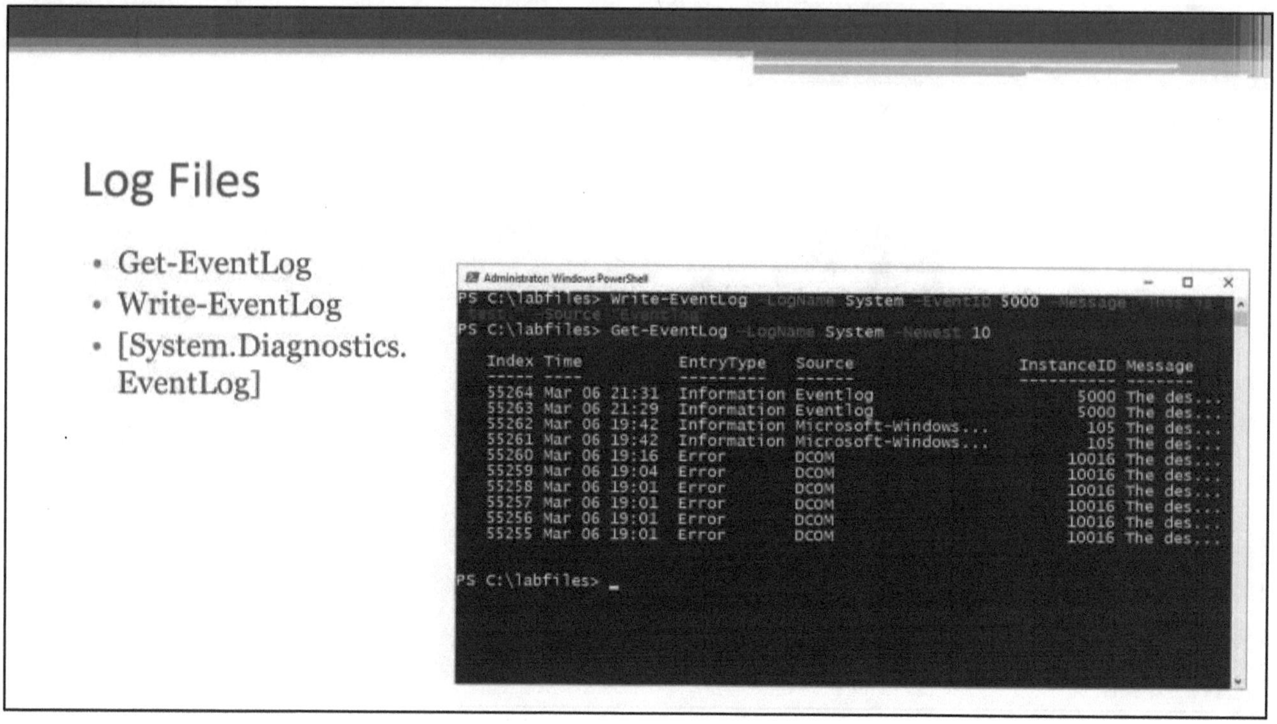

Information in Event Logs can be browsed by using the Get-EventLog cmdlet. A list of available log files can be seen with the –List parameter and other options allow you to specify what events you want to see from which log files (e.g. **Get-EventLog –LogName System –Newest 10**).

New events can be added to a log file with the Write-EventLog cmdlet. Parameters normally associated with a log entry can be specified such as category, eventid, source and message (e.g. **Write-EventLog –LogName Application –EventID 5000 –Message "This is a test." –Source "PS Entry"**). If the –Source option is not already linked with the –LogName option being used, the association will have to be created (e.g. **[System.Diagnostics.EventLog]::CreateEventSource("PS Entry","Application")**).

File System

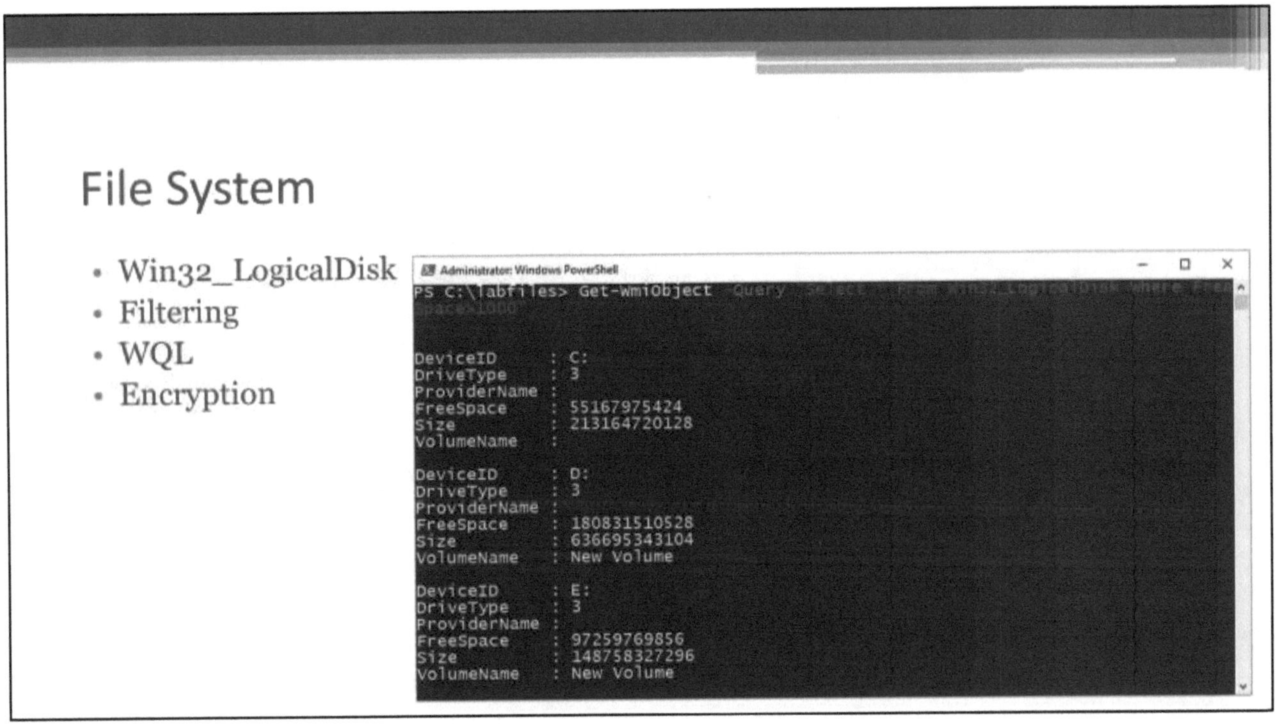

Browsing the file system can be done easily with the **Get-ChildItem** cmdlet. It has preconfigured aliases commonly used in the DOS and UNIX environments such as dir and ls. The creation or deletion of files and folders can be managed with the Add-Item and Remove-Item cmdlets. They also have commonly used aliases assigned to them. To view the information inside of a text file, the **Get-Content** cmdlet can be used.

Information about the drives on a system such as the total size or free space available is accessible with the Win32_LogicalDisk class. Filtering options can be used to specify which particular drive you need information on (e.g. **Get-WmiObject Win32_LogicalDisk –Filter "DeviceID='C:'"** or **Get-WmiObject Win32_LogicalDisk –Filter "FreeSpace > 2000000000"**). These statements can also be written using WMI Query Language (WQL) if preferred (e.g. **Get-WmiObject –Query "Select * From Win32_LogicalDisk Where DeviceID='C:'"** or **Get-WmiObject –Query "Select * From Win32_LogicalDisk Where FreeSpace>2000000000"**).

BitLocker encryption is can now be configured with the **Enable-BitLocker** cmdlet. Installed and removable drives can be encrypted using the computer's Trusted Platform Module (TPM), a smart-card or password. By default a TPM chip is required but this can be changed using group policy settings. Before BitLocker is enabled for any drive, a method of recovering the encryption keys should be planned and implemented.

Network Adapter

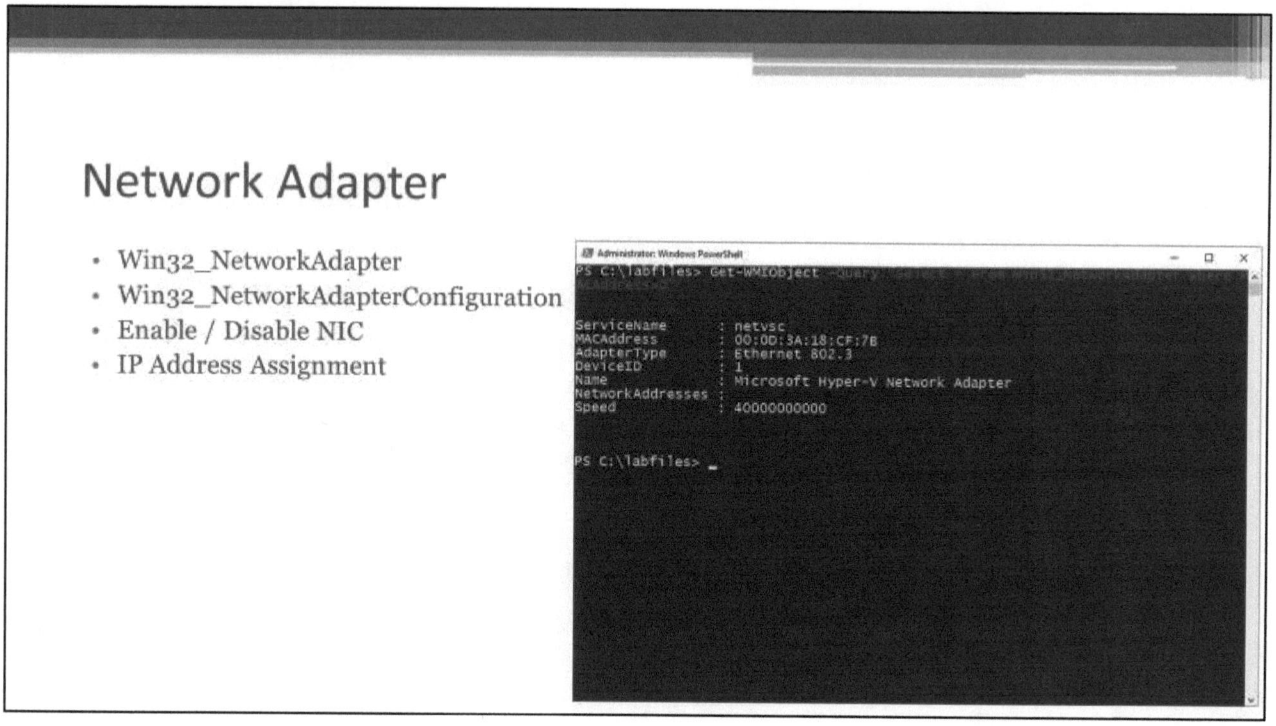

For information about network adapters, use the **Get-WmiObject Win32_NetworkAdapterConfiguration** command. Details about the card such as adapter speed, IP address and DHCP Settings will be displayed. MAC address information is accessible with the Win32_NetworkAdapter class. Assigning new settings for static IP configuration, enabling DHCP or enabling/disabling an adapter can be done with the Win32_NetworkAdapterConfiguration class.

Here is a method that can be used to enable or disable a network adapter by taking advantage of variable assignments:

$NetworkAdapter=Get-WmiObject –Q "Select * From Win32_NetworkAdapter Where AdapterType like '%ethernet%'"
$NetworkAdapter.Disable()
$NetWorkAdapter.Enable()

A similar method can be used to configure a network card for static or dynamic IP configuration:

$NetworkConfig=Get-WmiObject –Q "Select * From Win32_NetworkAdapterConfiguration Where DNSDomain='Contoso.com'"
$NetworkConfig.EnableStatic("192.168.10.140","255.255.255.0");
$NetworkConfig.SetGateways("192.168.10.100");
$NetworkConfig.SetDNSServerSearchOrder("192.168.10.100")
$NetworkConfig.EnableDHCP(); $NetworkConfig.SetDNSServerSearchOrder()

Lesson 5: Administering Network Resources

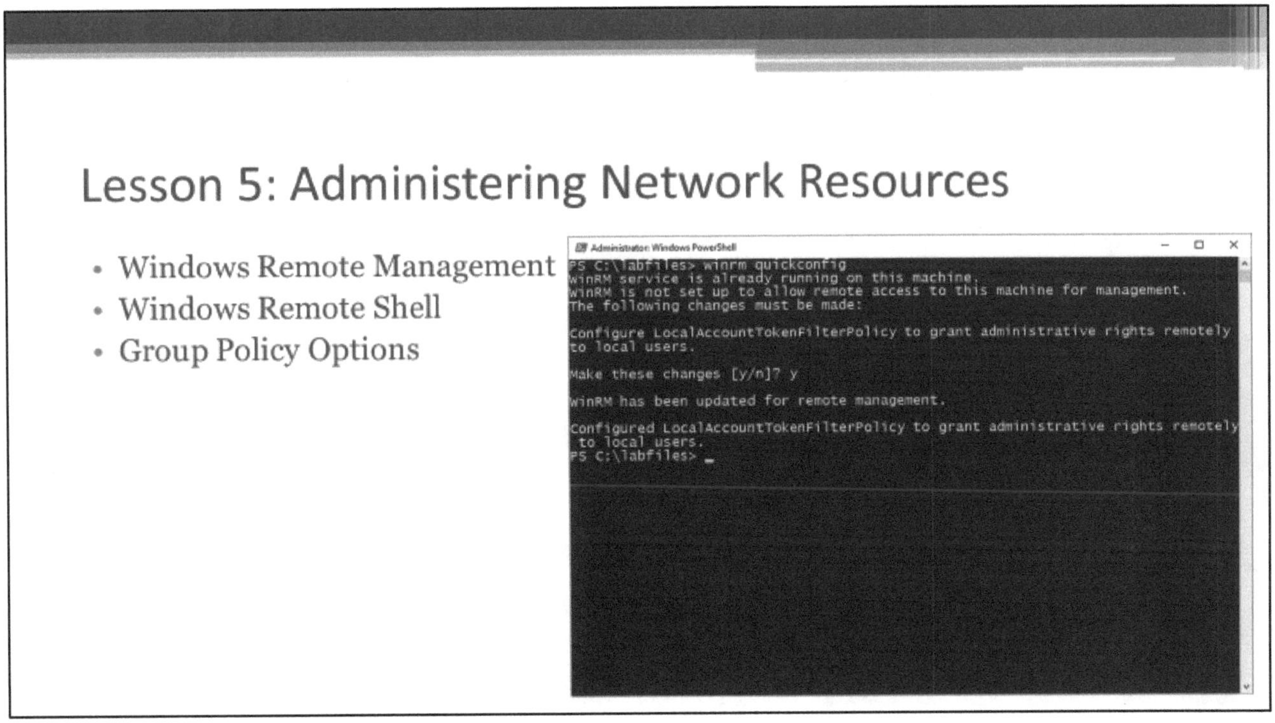

All the information available about a system locally using PowerShell cmdlets is accessible remotely using Windows Remote Management. This service can be configured from the command-line with the **winrm.exe** command (e.g. **winrm quickconfig**). It will automatically configure the appropriate firewall exceptions for the service to work properly. After the setup, you will be able to view and change remote computer settings. When combined with PowerShell scripting this allows for convenient remote administration of desktop components and features. **Enable-PSRemoting** is the PowerShell cmdlet used to enable remoting.

The Windows Remote Shell command (winrs.exe) allows you to execute PowerShell cmdlets, scripts and functions on remote systems as if you were sitting at the computer (e.g. **winrs –r:computer1 netstat –an**). The command allows different authentication credentials to be provided when performing these operations. Invoke-Command can be used to perform the same operations using PowerShell. A single command can be used to run processes on multiple computers (e.g. **Invoke-Command –Computer Computer1, Computer2 –Scriptblock {netstat –an}**)

Windows Remote Management and Windows Remote Shell can be centrally managed through Group Policy options. Because of the control this feature provides, it should be disabled unless an administrative argument can be made for using it.

Demonstration / Exercise 4

Note: Make sure your PowerShell console or ISE has administrator privileges when doing exercises. Any text editor can be used to create or edit the scripts. Students should be encouraged to use a popular PowerShell ISE or another GUI based scripting environment.

Demonstration: Show students how to get information about local and remote computer resources. Complete lessons 4 and 5 together before doing exercise 4. Perform exercise 4 before moving on to lesson 6.

Lesson 6: Connect to Microsoft Azure with PowerShell

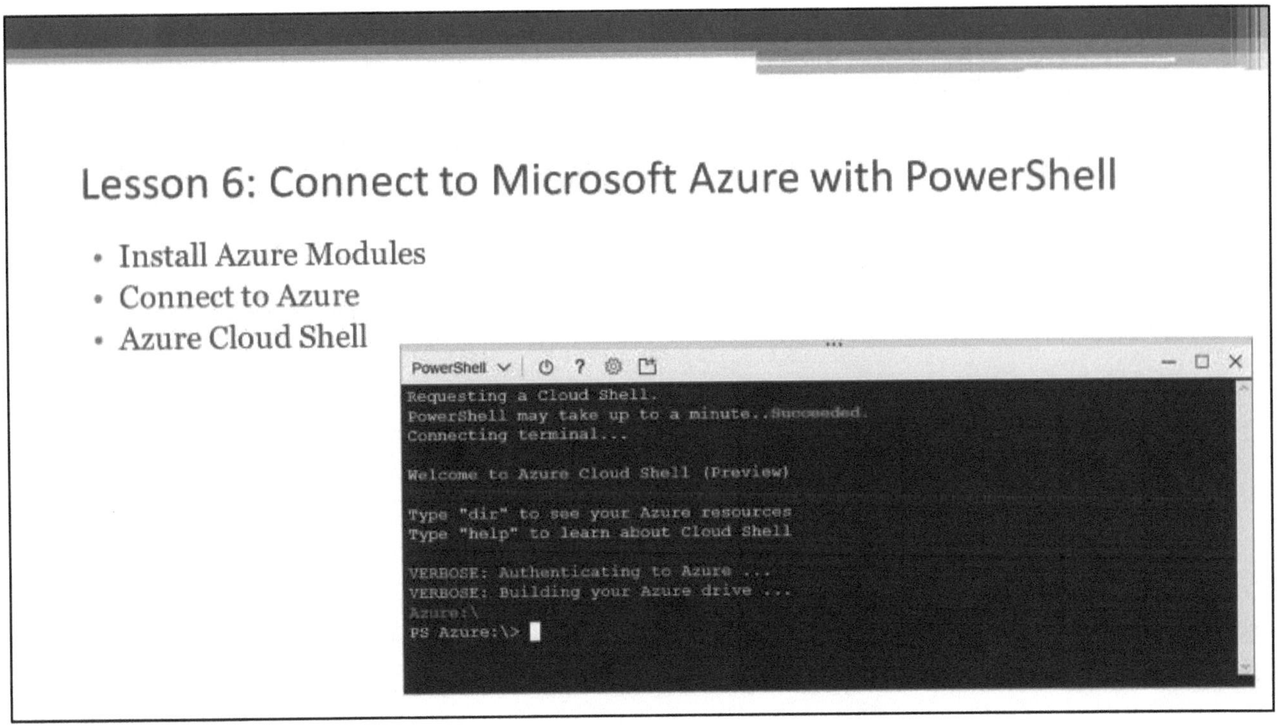

The tools and services in Microsoft Azure allow anyone to use IT resources in a cloud environment. Azure PowerShell makes all of these resources available from the command-line or in scripts. Getting started with Azure PowerShell is made simple with the use of cmdlets provided in the different Azure modules.

In this lesson, we will demonstrate how to get started using PowerShell to work in an Azure environment. You will learn to use Azure PowerShell to:

1. Install the Azure Modules
2. Connect to Azure with your subscription
3. Use the Azure Cloud Shell

The Azure web-site (http://www.azure.com) provides access to the Azure Portal and information that will teach you to graphically manage resources you create in your subscription and estimate the cost of using them.

Install the Azure Modules

Two modules must be installed to properly manage all Azure resources and these are discussed below. A module may have multiple versions available which may have important differences in how they are run. Knowing the version used by other people on your team and using the same one will help to avoid compatibility problems when using the same scripts and managing the same resources. Different versions of a module can be specified and run in different PowerShell consoles (e.g. **Import-Module AzureRM –RequiredVersion 2.3.0***). More detailed documentation can be reviewed on the PowerShell Gallery (http://www.powershellgallery.com), GitHub (http://www.github.com/azure) or the Azure web-site. Before going to the next section, install and import both Azure modules as described below.

> **Azure Resource Manager**: Most of the cmdlets used to manage Azure resources are a part of the Azure Resource Manager module. This module can be installed by running **Install-Module AzureRM** and cmdlets are then imported with the **Import-Module AzureRM** command. When installing a module, repositories from the PowerShell Gallery are used. The Set-PSRepository cmdlet can be used to configure the gallery as a trusted resource. (e.g. **Set-PSRepository –Name "PSGallery" –SourceLocation "https://www.powershellgallery.com/api/v2/" –InstallationPolicy Trusted**).

> **Azure Service Management**: To manage task scheduling, storage accounts and hosted services, the Azure Service Management cmdlets can be used. This module is installed by running the **Install-Module Azure** command and the cmdlets are imported with the **Import-Module Azure** command.

Connect to Azure

The Azure Resource Manager cmdlets (AzureRM) are used to login to Azure and manage your subscription. A list of all AzureRM cmdlets can be retrieved by running **Get-Command –Module AzureRM**. You login to Azure by running **Login-AzureRmAccount** and providing your login credentials. Next, specify the subscription you will be using. If you only have one subscription, running the command **Get-AzureRmSubscription -SubscriptionName <SubscriptionName> | Select-AzureRmSubscription** is an easy way to accomplish this.

Azure Cloud Shell

The easiest option for using PowerShell cmdlets in Azure is to use the browser-based Azure Cloud Shell. From the Azure Portal, an authenticated shell can be generated that uses a Bash or PowerShell console to manage objects. Clicking the Cloud Shell icon ( ) will generate a new console from which multiple session windows can be created and used separately.

The console will use the credentials of the Azure Portal. Because the Azure modules are already installed and credentials are already specified, you may start working immediately without needing any additional configuration. Both the Bash and PowerShell consoles support Azure CLI and Python.

Creating and Managing resources in Azure

Before deciding what objects to create in Azure, you should consider how they will be managed. Resources can be deployed using the Classic or Resource Group model:

The classic model requires that each resource (e.g. storage space, network components) needed for a solution (e.g. virtual machine) be deployed individually and in the right order. Deleting a solution requires that each resource be deleted individually. A default resource group is assigned to classic solutions when they are created.

The Resource Manager model uses resource groups to organize resources in a solution. They can be deployed, managed and deleted as a group. Permissions assigned to a resource group automatically apply to all objects in it. We will focus on using the Resource Manager model in this lesson.

Each resource group is assigned a location (e.g. East US) at the time of its creation. A common strategy when working with temporary objects is to assign them to a temporary resource group. Simply deleting the resource group will delete all objects created in it. This strategy will be used for our lab exercises.

Another important Azure resource is a storage account. This allows you to create different types of cloud storage components such as blobs, queues, disks, data lakes, etc. These are used to store data in different formats and disks used for backups and virtual machines. Storage account names must be unique throughout the Microsoft Azure environment.

The 50331D, 55066A, 55069A, 55264A, 55224X, 55315A, 55316A and 55318A courses from Microsoft Learning include Azure PowerShell scripts that can be used to create virtual machines in Microsoft Azure. That process is replicated in one of the exercises so you will be able to use it.

Demonstration / Exercise 5

Note: Make sure your PowerShell console or ISE has administrator privileges when doing exercises. Any text editor can be used to create or edit the scripts. Students should be encouraged to use a popular PowerShell ISE or another GUI based scripting environment.

Demonstration: Show students how to use the local machine PowerShell ISE / Console or Azure Cloud Shell to create an Azure virtual machine. The ***AzureClientVM.ps1*** (local machine) and ***AzureSetup_cs.ps1*** (Azure Cloud Shell) scripts will be kept current in case there are changes to the PowerShell cmdlets used for this exercise.

Resolve PowerShell Scripting Problems

When using PowerShell, technology professionals may use it for a wide range of tasks. Problems encountered when executing code interactively or by using a script will often be related to configuration, syntax or permission issues. In this section, we will look at some of those issues and how they might be resolved.

You want to backup components in Microsoft Azure so they can be recreated later for another project. Can this be done with PowerShell?
PowerShell cmdlets can be used to export Azure information and then recreate them at a later date. (**Export-AzureRmResourceGroup** command can be used to save the configuration of objects in Azure to json files and **New-AzureRmResourceGroupDeployment** can be used to recreate them from these same files.)

You are unable to get examples of how to use a cmdlet by running the command Get-Help Get-AzureRmStorageAccount –Examples. The command works but you do not have the help files installed. How can you get them?
Run the command **Update-Help**. If you know the specific module name, use **Update-Help –Module [ModuleName]**

Your network has Windows 10 and 8 devices. PowerShell is installed and configured on all of them. You realize that some of the commands you use on the Windows 10 systems are not available on Windows 8. What could be the cause of this problem?
They are probably using different versions of PowerShell. Upgrade the systems so they all use the same version of PowerShell.

Some of the cmdlets used in your PowerShell scripts can make significant changes to a domain computer. How can you add prompts to the code that require confirmation before a cmdlet is executed?
All PowerShell comdlets can use the –Confirm parameter. It will prompt the user to carry out the operation, stop it or suspend it.

You have been having trouble with some PowerShell cmdlets on your system. They are from different modules and reinstalling them has not fixed the problem. How can you get them to work again?
Deleting the specific module folders from the modules directory on your computer (e.g. C:\program Files\WindowsPowerShell\Modules) and reinstalling the modules is sometimes necessary.

To simplify the testing of scripts, a member of your team has changed the command associated with PS1 files from notepad.exe to powershell.exe and recommended that this be done on all systems. How might such a change affect script security?
The default configuration of not associating PS1 files with powershell.exe was deliberately done as a security measure to prevent the accidental execution of scripts, whether they have malicious code or not. While this association might be changed on testing and development computers, it should not be done on production systems.

After testing a script on the local computer, you are unsuccessful when trying to remotely execute it on another computer. After verifying that the remote system has PowerShell installed, what else should you check?
Make sure that Windows Remote Management is configured and that the PowerShell execution policy allows remote execution of scripts.

A new member of your team has recommended that the PowerShell execution policy be set to unrestricted because the network does not have a Certificate Authority and scripts will be used extensively for administrative purposes. What are the pitfalls of this recommendation and how can you solve this problem without compromising security?
An execution policy of unrestricted allows scripts to be executed without verification of their source. The Makecert.exe command can be used to generate a certificate that can be used for signing scripts. This can be done without a CA but still be deployed as trusted through Active Directory Group Policy settings.

When you try to execute the Set-ExecutionPolicy cmdlet on a computer, it fails because you are unable to make the needed changes to the registry. How can you solve this problem?
Open a new shell with Administrative credentials and run the command again.

Some of the professionals on your team are new to PowerShell and need help to understand the functionality of some common cmdlets used for desktop management. How can they get this kind of information directly from the shell?

They can use the Get-Help cmdlet, the Help command or the Help parameter (e.g. **Get-Help Get-Service** or **Help Get-Service** or **Get-Service -?**). They can also use the help files that provide information about different features and options in PowerShell. A list of the help files is accessible by running the command **Get-Help *about*.**

Some of your support staff are complaining about the names used for scripts and variables created for Microsoft Azure jobs. What rules can be followed to make sure that appropriate names are assigned to these objects?

For scripts, always use names that are descriptive, use pascal casing for the names and use the Verb-Noun syntax used with cmdlets. For variables, use descriptive names and camel casing for the names.

A new member of your team with a strong VBScript background is recommending it as a scripting solution instead of PowerShell. They do not see any advantages to PowerShell and insist they would have to re-write existing scripts if the company moved to that environment. How can you respond to these statements?

PowerShell will have performance gains vs. VBScripts in some situations and offers more security options for protecting computer systems. It is also easier to learn. Existing VBScripts will not have to be re-written to run in a PowerShell environment.

Your network will be migrating from a VBScript to a PowerShell environment. Before the changes are made your department manager wants you to create a report detailing the advantages of VBScript over PowerShell. What are some of the details you would include in such a report?

VBScript is supported on more versions of Windows than PowerShell and does not need any components to be installed for it to work. All currently supported versions of Windows have PowerShell installed by default.

You need to use Azure PowerShell to work with storage components on the platform but you get an error message that the command is not available. Other Azure cmdlets work fine. What can you do to solve this problem?

Find out the module that the cmdlet belongs to and install it. Not all Azure PowerShell cmdlets are included with AzureRM.

Review – PowerShell for Technology Professionals

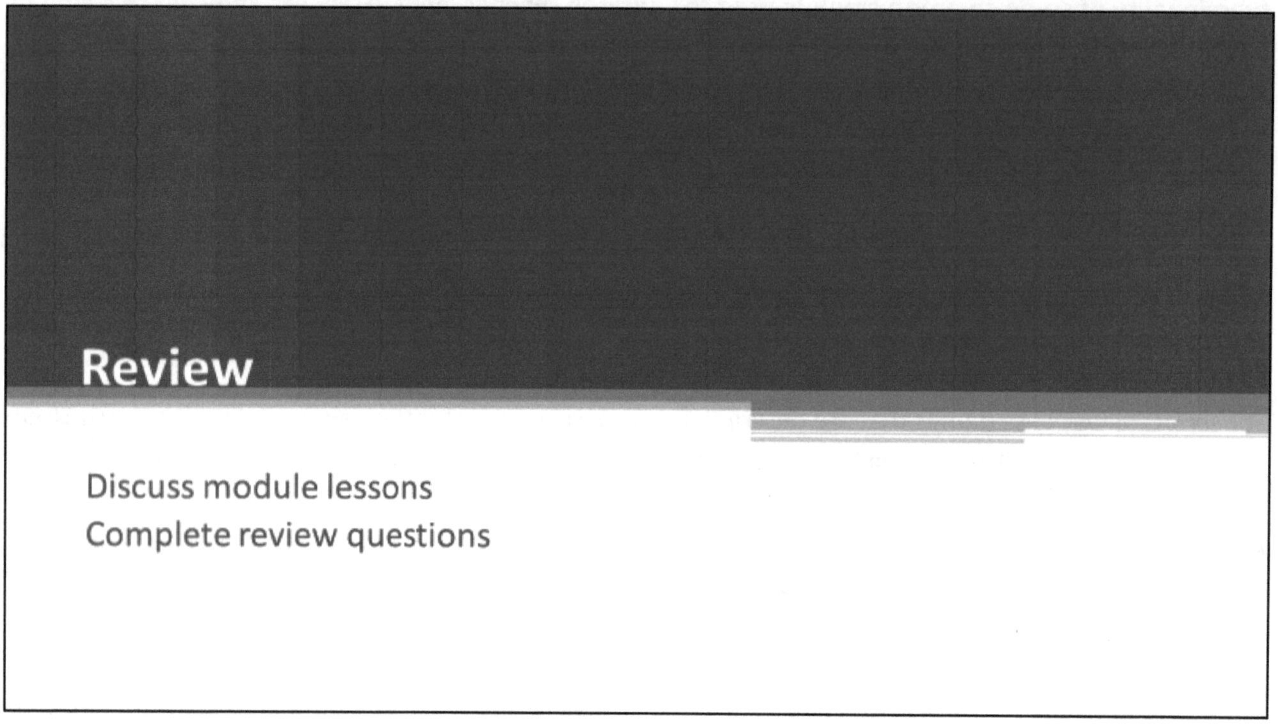

1. What cmdlet is used to enable the use of scripts in PowerShell?

2. What service must be enabled to run PowerShell scripts on a remote system?

3. What tool can be used to create a certificate for a script?

4. True or False. The data type of a variable can be changed dynamically in a script.

5. What cmdlet can be used to list all PowerShell commands?

6. What command could you use to get helpful information about using the New-Alias cmdlet?

7. How can you send the output from the Get-ChildItem cmdlet to a new file named Info.txt?

8. How can you append the errors generated by the Get-Help -f command to a file named Errors.txt?

9. True or False. CMD.exe and UNIX commands can be executed from a PowerShell shell.

10. How can you capture a list of filenames that have a TXT extension?

11. How can you generate a list of all the aliases configured in a Shell?

12. True or False. Script debugging can only be done manually from the command-line or a text editor.

13. How can you install Azure PowerShell on a desktop?

Lab B: PowerShell for Technology Professionals (2 – 3 hours)

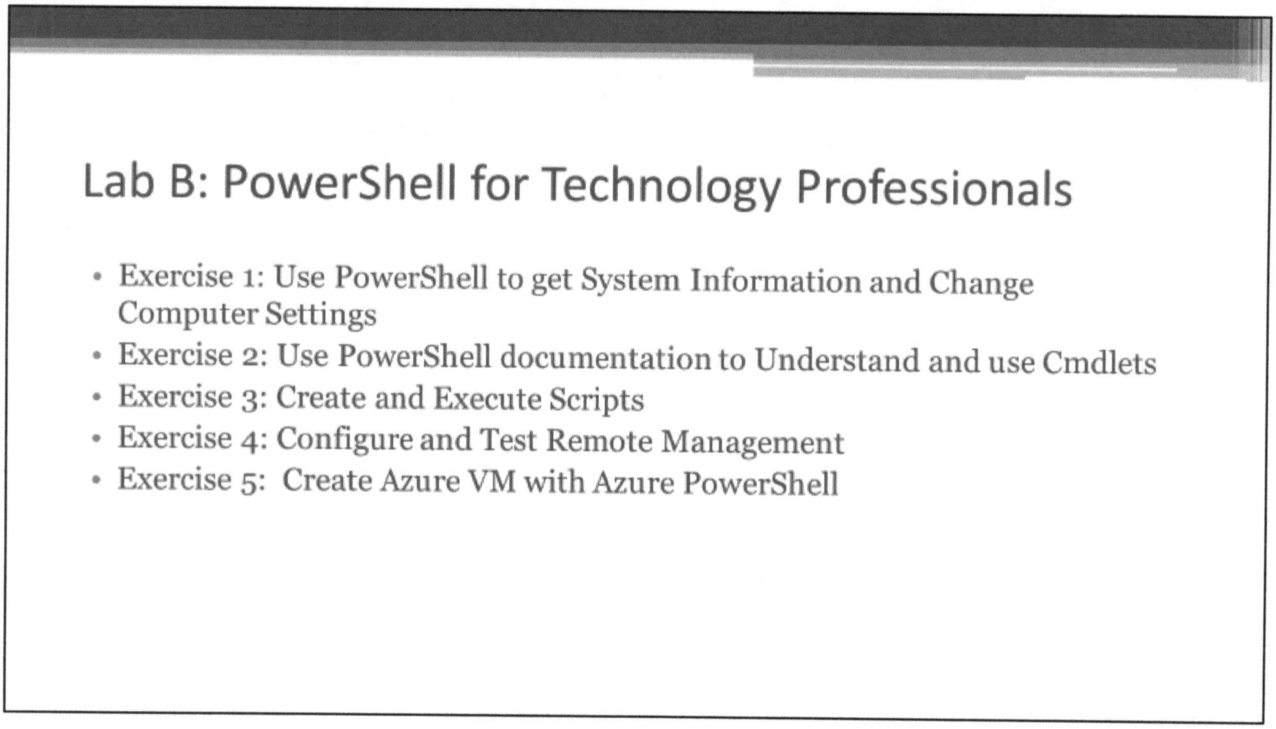

Overview: Exercises will teach students how to run PowerShell commands interactively and from scripts. You will perform tasks to configure PowerShell security options and setup remote management using PowerShell. You may also create a virtual machine in Azure if you have a subscription to it or sign up for a trial one. References to NYC-DC1 or Student10 in the labs should be replaced with your own computer names. If you do not have a test computer to work on, do Exercise 5 first in order to setup one in Azure. Some exercises require two computers, so you may decide to create two virtual machines in Azure.

Estimated Time to complete this lab is 120 - 180 minutes.

Exercise 1: Use PowerShell to get System Information and Change Computer Settings

1. Login to Computer1 as Contoso\Admin1.
2. Open the PowerShell console as an Administrator by clicking Start > All Programs > Accessories > Windows PowerShell, Right Click the Windows PowerShell icon and click "Run as administrator". Click Yes in the "User Account Control" window.
3. Run the following commands and notice the results:
    - Get-Command
    - Get-Command | Measure-Object

- Get-Help | More
- Get-Help Get-WMIObject –Detail | More
- Get-Help *about*
- Get-Help About_Scripts | More
- CD ~; Cls; Dir
- Get-Alias CD; Get-Alias CLS; Get-Alias Dir
- Get-WmiObject Win32_NetworkAdapterConfiguration
- Get-WmiObject –Query “Select * From Win32_NetworkAdapterConfiguration”
- Get-WmiObject –Query “Select * From Win32_NetworkAdapterConfiguration Where DNSDomain=’Contoso.com’”
- Get-WmiObject –Query “Select * From Win32_NetworkAdapterConfiguration” | Format-Table IPAddress,MACAddress,DNSDomain
- Get-WmiObject Win32_ComputerSystem | Format-Table Name,Domain,TotalPhysicalMemory
- Set-Location HKLM:\Software; Dir; Set-Location C:
- New-Item –Path c:\temp\tmp –type directory
- Set-Location c:\temp\tmp
- New-Item –Path c:\temp\tmp\test.txt –type file
- Set-Alias Copy-Con Set-Content
- Copy-Con test.txt “This is a test”
- Get-Content test.txt
- Get-Process
- Get-Process | Select-Object ProcessName, ID | Export-Csv test1.csv
- Get-Process | Format-Table ProcessName, ID | Export-Csv test2.csv
- Get-Content test1.csv, test2.csv | More
- Invoke-Command –Scriptblock {notepad}
- Start-Process ‘c:\windows\system32\notepad.exe’
- Get-Process notepad
- Get-Process notepad | Stop-Process
- Get-PSDrive F
- Get-BitLockerVolume
- $Password = ConvertTo-SecureString "ABCDEFG123" -AsPlainText -Force
- Enable-BitLocker -MountPoint "F:" -EncryptionMethod Aes256 -UsedSpaceOnly -Password $Password -PasswordProtector
- Get-BitLockerVolume -MountingPoint 'F:' | Disable-Bitlocker
- Restart-Computer

Exercise 2: Use PowerShell Documentation to Understand and use Cmdlets

Note: For each of the following steps, use the information in Lessons 2 – 4 and Exercise 1 to choose the correct command for each task. Unless otherwise stated, each task must be completed with a PowerShell cmdlet. If a task changes the computer configuration, verify that the operation completed successfully.

1. Get a list of all the Set- commands.

2. Get detailed information about the Stop-Service cmdlet.

3. Stop the Spooler service and then restart it.

4. Get a list of all active processes running on the computer

5. Get the process ID number for any process with the name iexplore (To perform this step, start an Internet Explorer session if you do not already have one running.)

6. Stop an Internet Explorer session using the process ID number assigned to it.

7. Assign the string "This is a test" to a variable named $String1.

8. Assign the string " of concatenation." to a variable named $String2.

9. Concatenate $String1 and $String2 as a new variable named $String3 and print its value to the screen.

10. Do a folder and file listing in your home directory and send the data to a text file named C:\Temp\Dir.txt.

11. Peform a directory listing of the C:\Users folder and append the information to the C:\Temp\Dir.txt file.

12. Perform a directory listing on the fictitious folder named C:\Temporary and redirect the error message to C:\Temp\Errors.txt.

13. Perform a query to get information about all logical drives on the computers.

14. Display the logical drives in a **list**, showing only their name, size and freespace.

15. Display the logical drives in a **table**, showing only their name,size and freespace. Do not include drives with zero drive space.

16. Perform a query to get the amount of free disk space on drive C:

17. Create a folder named C;\Logs

18. Create two files named C:\Logs\Errors.log and C:\Logs\Errors2.log

19. Add two lines to the C:\Logs\Errors.log file that say "Error Number 1" and "Error Number 2".

20. Delete the C:\Logs\Errors2.log file.

21. Rename the C:\Logs\Errors.log file to C:\Logs\Errors.dat.

22. View the last 5 entries in the Application Event Log.

23. Record the last 50 entries from the System Event Log to a file named C:\Log\System.log

24. Create a new Source for the Event Viewer Application Log named "Admin Script".

25. Add an entry to the Application log with an EventID of 5500, a Message of "The Task Completed Successfully." and a Source of "Admin Script".

26. Disable the Network Adapter.

27. Enable the Network Adapter.

28. Assign a static IP address of 192.168.10.50 to the network card.

29. Configure the network adapter to use DHCP.

30. Add a folder to the HKEY_CURRENT_USER registry hive named PSLogic, verify that it was added correctly and then remove it.

Exercise 3: Create and Execute Scripts

1. Open Notepad and add the following lines to the text file:
    - # This script will copy all events from the Windows PowerShell Log to the C:\Log\PS.log file
    - # It will clear events from the Windows Power Shell Log after getting confirmation from the user.
    - Get-EventLog "Windows PowerShell" > C:\Log\PS.log
    - Clear-Eventlog "Windows PowerShell" –Confirm
2. Save the file with the name C:\Temp\backupPowerShellLog.ps1. Make sure it does not get saved with a txt extension.
3. Open a PowerShell console with administrative credentials.
4. Run the cd \temp command to go to the C:\Temp folder.
5. Run the command backupPowerShellLog.ps1. Make a note of the error message. You must specify the full or relative path of the script (e.g. c:\temp\backupPowerShellLog.ps1).
6. Run the command .\backupPowerShellLog.ps1. Make a note of the error message. The execution policy prevents the running of scripts by default.
7. Run the command Get-Executionpolicy to verify that the execution policy is set to Restricted.
8. Run the command Set-Executionpolicy Unrestricted.
9. Run the command .\backupPowerShellLog.ps1. It should run successfully this time. Confirm the deletion of messages in the Windows PowerShell log when asked to do so.
10. Verify that the C:\Log\PS.log file has the log information and that the Windows PowerShell Event Log is empty (**Get-Content C:\Log\PS.log; Get-EventLog –List**).

Exercise 4: Configure and Test Remote Management

1. Login to NYC-DC1 as Contoso\Administrator.
2. Open the PowerShell console as an Administrator by clicking Start > All Programs > Accessories > Windows PowerShell, right click the Windows PowerShell icon and click "Run as administrator".
3. Ping Computer1 to verify connectivity. If necessary, temporarily stop the Windows Firewall service on Computer1 (Run Services.msc and stop the Windows Firewall service).
4. Run the command: **Get-WmiObject –computer Computer10 Win32_ComputerSystem**.
5. Make a note of the error message. You cannot perform remote management because WinRM has not been configured on the Windows client.
6. Login to Computer1 as Contoso\Administrator.
7. Open the PowerShell console as an Administrator by clicking Start > All Programs > Accessories > Windows PowerShell, right click the Windows PowerShell icon and click "Run as administrator".
8. Run the command: **winrm quickconfig**. At each prompt, type Y and press Enter to accept the changes being made. This will configure remote management on the system.
9. From NYC-DC1, run the command: **Get-WmiObject –computer Computer10 Win32_ComputerSystem** again. It should be successful this time. If not successful, restart the "Windows Remote Management" service on Computer1 and try again.
10. Using the information from Exercise 1 and 2, use PowerShell cmdlets to perform the following tasks on Computer1 while logged into NYC-DC1:
    - Get the MAC Address of the network card
    - Get a list of all running services
    - Find out how much RAM is on the computer
    - Find out how much free space is available on the C: drive
    - View the last 10 entries in the Event Viewer Security Log
11. From NYC-DC1, execute the following commands to test the Windows Remote Shell command:
    - Invoke-Command -Computer Computer10 –ScriptBlock {hostname}
    - Invoke-Command -Computer Computer10 –ScriptBlock {ipconfig /all}
    - Invoke-Command -Computer Computer10 –ScriptBlock {nbtstat –n}
    - Invoke-Command -Computer Computer10 –ScriptBlock {netstat –n}
    - Invoke-Command -Computer Computer10 -ScriptBlock {md c:\temp2}
    - Invoke-Command -Computer Computer10 -ScriptBlock {net share temp2=c:\temp2}
    - Invoke-Command -Computer Computer10 -ScriptBlock {net share}

Exercise 5: Create an Azure VM with Azure PowerShell

Note: The script that corresponds to this exercise is **AzureClientVM.ps1** and should be located in the C:\Labfiles folder. If you prefer to use the Azure Cloud Shell, use **AzureSetup_cs.ps1** instead. Run all commands with a PowerShell Administrator console. Make sure you have your Azure Subscription information and a reliable Internet connection before proceeding.

1. Configure Objects & Variables
   a) Set-StrictMode -Version 2.0
   b) $SubscriptionName = "Azure Pass" # Replace with the name of your Azure Subscription
   c) $WorkFolder = "C:\Labfiles\"
   d) Set-Location $WorkFolder
   e) $PW = Write-Output 'Pa$$w0rdPa$$w0rd' | ConvertTo-SecureString -AsPlainText -Force # Password for Administrator account
   f) $AdminCred = New-Object System.Management.Automation.PSCredential("Adminz",$PW) # Login credentials for Administrator account
   g) $Location = "EASTUS"
   h) $VMName = "vm55264srv"
   i) $NamePrefix = "in" + (Get-Date -Format "HHmmss") # Replace "in" with your initials. Date information is added in this example to help make the names unique
   j) $ResourceGroupName = $NamePrefix.ToLower() + "rg"
   k) $StorageAccountName = $NamePrefix.ToLower() + "sa" # Must be lower case
   l) $SAShare = "55264a" # Must be lower case

2. Login to Azure
   a) Connect-AzureRMAccount
   b) $Subscription = Get-AzureRmSubscription -SubscriptionName $SubscriptionName | Set-AzureRMContext

3. Create Resource Group, Storage Account & Storage Account Share
   a) New-AzureRMResourceGroup -Name $ResourceGroupName -Location $Location
   b) New-AzureRMStorageAccount -ResourceGroupName $ResourceGroupName -Name $StorageAccountName -Location $Location -Type Standard_RAGRS
   c) $StorageAccountKey = (Get-AzureRmStorageAccountKey -ResourceGroupName $ResourceGroupName -Name $StorageAccountName)[0].Value
   d) $StorageAccountContext = New-AzureStorageContext -StorageAccountName $StorageAccountName -StorageAccountKey $StorageAccountKey
   e) New-AzureStorageContainer -Name $SAShare.ToLower() -Context $StorageAccountContext -Permission Container -Verbose
   f) Get-ChildItem $WorkFolder"55264CustomScriptExtension.ps1" | Set-AzureStorageBlobContent -Container $SAShare -Context $StorageAccountContext -Force

4. Create Network
   a) $NSGRule1 = New-AzureRmNetworkSecurityRuleConfig -Name "RDPRule" -Protocol Tcp -Direction Inbound -Priority 1001 -SourceAddressPrefix * -SourcePortRange * -DestinationAddressPrefix * -DestinationPortRange 3389 -Access Allow

b) $NSGRule2 = New-AzureRmNetworkSecurityRuleConfig -Name "MSSQLRule" -Protocol Tcp -Direction Inbound -Priority 1002 -SourceAddressPrefix * -SourcePortRange * -DestinationAddressPrefix * -DestinationPortRange 1433 -Access Allow
c) $NSGRule3 = New-AzureRmNetworkSecurityRuleConfig -Name "WinHTTP" -Protocol Tcp -Direction Inbound -Priority 1003 -SourceAddressPrefix * -SourcePortRange * -DestinationAddressPrefix * -DestinationPortRange 5985 -Access Allow
d) $NSGRule4 = New-AzureRmNetworkSecurityRuleConfig -Name "WinHTTPS" -Protocol Tcp -Direction Inbound -Priority 1004 -SourceAddressPrefix * -SourcePortRange * -DestinationAddressPrefix * -DestinationPortRange 5986 -Access Allow
e) $NSG1 = New-AzureRMNetworkSecurityGroup -ResourceGroupName $ResourceGroupName -Location $Location -Name "NSG1" -SecurityRules $NSGRule1,$NSGRule2,$NSGRule3,$NSGRule4 -Force
f) $Subnet1 = New-AzureRmVirtualNetworkSubnetConfig -Name "Subnet1" -AddressPrefix 192.168.10.0/24
g) $VirtualNetwork1 = New-AzureRmVirtualNetwork -Name "VirtualNetwork1" -ResourceGroupName $ResourceGroupName -Location $Location -AddressPrefix 192.168.0.0/16 -Subnet $Subnet1
h) $PublicIP1 = New-AzureRmPublicIpAddress -Name "PublicIP1" -ResourceGroupName $ResourceGroupName -Location $Location -SKU Standard -AllocationMethod "Static" -Force
i) $NIC1 = New-AzureRmNetworkInterface -Name "NIC1" -ResourceGroupName $ResourceGroupName -Location $Location -SubnetId $VirtualNetwork1.Subnets[0].Id -PublicIpAddressId $PublicIP1.Id -NetworkSecurityGroupId $NSG1.Id

5. Create SQL Server VM
   a) $PublisherName = "MicrosoftSQLServer"
   b) $Offer = "SQL2017-WS2016"
   c) $Skus = "SQLDEV"
   d) $VMSize = (Get-AzureRMVMSize -Location $Location | Where-Object {$_.Name -like "Standard_DS2*"})[0].Name
   e) $VM1 = New-AzureRmVMConfig -VMName $VMName -VMSize $VMSize
   f) $VM1 = Set-AzureRmVMOperatingSystem -VM $VM1 -Windows -ComputerName $VMName -Credential $AdminCred -WinRMHttp -ProvisionVMAgent -EnableAutoUpdate
   g) $VM1 = Set-AzureRmVMSourceImage -VM $VM1 -PublisherName $PublisherName -Offer $Offer -Skus $Skus -Version "latest"
   h) $VM1 = Add-AzureRMVMNetworkInterface -VM $VM1 -ID $NIC1.Id
   i) $VHDURI1 = (Get-AzureRMStorageAccount -ResourceGroupName $ResourceGroupName -Name $StorageAccountName).PrimaryEndPoints.Blob.ToString() + "vhdsrv/VHDSRV1.vhd"
   j) $VM1 = Set-AzureRmVMOSDisk -VM $VM1 -Name "VHDSRV1" -VHDURI $VHDURI1 -CreateOption FromImage
   k) New-AzureRmVM -ResourceGroupName $ResourceGroupName -Location $Location -VM $VM1 -Verbose
   l) Start-AzureRMVM -Name $VMName -ResourceGroupName $ResourceGroupName
   m) Set-AzureRmVMCustomScriptExtension -Name "Microsoft.Compute" -TypeHandlerVersion "1.1" -FileName "55264CustomScriptExtension.ps1" -Run "55264CustomScriptExtension.ps1" -ForceRerun $(New-Guid).Guid -ContainerName $SAShare -ResourceGroupName $ResourceGroupName -VMName $VMName -Location $Location -StorageAccountName $StorageAccountName -StorageAccountKey $StorageAccountKey

n) $PublicIPAddress1 = Get-AzureRmPublicIpAddress -Name "PublicIP1" -ResourceGroupName $ResourceGroupName
o) Write-Output "The virtual machine has been created. Login as Adminz by using Remote Desktop Connection to connect to its Public IP address. The IP Address will be added to your trusted hosts list."
p) Write-Output "Public IP Address for $VMName is: " $PublicIPAddress1.IPAddress
q) Set-Item wsman:localhost\client\trustedhosts -value $PublicIPAddress1.IPAddress -Concatenate -Force
r) $AdminUser=$PublicIPAddress1.IPAddress + "`\Adminz"
s) $AdminCred2 = New-Object System.Management.Automation.PSCredential($AdminUser,$PW) # Login credentials for Administrator account
t) $PSSession = New-PSSession -ComputerName $PublicIPAddress1.IPAddress -Credential $AdminCred2
u) Copy-Item -ToSession $PSSession -Path $WorkFolder"55264A-ENU_PowerShellSetup.zip" -Destination "C:\Labfiles\" -Recurse -Force -ErrorAction Continue
v) Copy-Item -ToSession $PSSession -Path $WorkFolder"55264ConfigAZVM.ps1" -Destination "C:\Labfiles\" -Recurse -Force -ErrorAction Continue
w) Copy-Item -ToSession $PSSession -Path $WorkFolder"*.xml" -Destination "C:\Labfiles\" -Recurse -Force -ErrorAction Continue
x) Invoke-Command -ComputerName $PublicIPAddress1.IPAddress -Credential $AdminCred2 -ScriptBlock {Expand-Archive -LiteralPath "C:\Labfiles\55264A-ENU_PowerShellSetup.zip" -DestinationPath "C:\Labfiles" -Force}
y) Invoke-Command -ComputerName $PublicIPAddress1.IPAddress -Credential $AdminCred2 -ScriptBlock {C:\Labfiles\55264ConfigAZVM.ps1}
z) Remove-PSSession $PSSession

6. Remove Azure resources after you are finished using them. Active resources incur costs for your subscription.
   a) Remove-AzureRMResourceGroup -Name $ResourceGroupName -Verbose -Force
   b) Clear-Item WSMan:\localhost\Client\TrustedHosts -Force